U.S. Marine Corps Physical Readiness Training for Combat

Plus

U.S. Marine Corps Iraq: An Introduction to the Country and People

Department of Defense

Pentagon
Publishing

UNITED STATES MARINE CORPS
Marine Corps Combat Development Command
Quantico, Virginia 22134-5001

29 January 1988

FLEET MARINE FORCE REFERENCE PUBLICATION..........0-1B

MARINE PHYSICAL READINESS TRAINING FOR COMBAT

1. PURPOSE
Fleet Marine Force Reference Publication (FMFRP) 0-lMarine Physical Readiness Training for Combat is intended for use by all Marines. It provides the information and references necessary to establish and conduct physical conditioning programs to prepare Marines for the physical demands of combat.

2. APPLICABILITY
Although not directive, this handbook is intended for use as a reference by all Marines in developing unit physical readiness training for combat. This FMFRP does not establish policy for the conduct of physical fitness training. Refer to Marine Corps Order 6300.3_ for applicable Marine Corps policy on physical fitness. It may also be used as a basis for instruction conducted by unit leaders or as a reference for instruction or training.

3. RECOMMENDATIONS
To improve this manual, the MAGTF Warfighting Center seeks comments and recommendations from the user. Officers and noncommissioned officers with constructive comments are encouraged to respond. Lack of response will be considered as concurrence. The User Suggestion Form which follows can be detached and forwarded to - -

Commanding General
Marine Corps Combat Development Command (WF 06)
Quantico, Virginia 22134-5001

4. DISTRIBUTION
Initial distribution of this FMFRP 0-1B is shown below. Addressees on this list should ensure that appropriate subordinate organizations are provided copies.

JOHN I. HOPKINS
Brigadier General, U.S. Marine Corps
Assistant Chief of Staff, Warfighting
Marine Corps Combat Development Command
Quantico, Virginia

DISTRIBUTION: TDJ

From:

To: Commanding General, Marine Corps Combat Development Command
(WF 06), Quantico, Virginia 22134-5001

Subj: FMFRP 0-1B, MARINE PHYSICAL READINESS TRAINING FOR COMBAT, RECOMMENDATIONS CONCERNING

1. In accordance with the Promulgation Page to FMFRP 0-1B, which invites individuals to submit suggestions concerning this FMFRP directly to the above addressee, the following unclassified recommendation is forwarded:

Page_________Article/Para No._________Line No._________Figure/Table No._________

Nature of Change:_____Add_____Delete_____Change_____Correct

2. Proposed New Verbatim Text: (Verbatim, double spaced; continue on additional pages as necessary.)

3. Justification/Source: (Need not be double spaced.)

NOTE: Only one recommendation per page.

PREFACE

MARINE PHYSICAL READINESS TRAINING FOR COMBAT

Chapter 1
PHYSICAL READINESS LEADERSHIP
Section I. Role of Physical Fitness in Combat

1101. PURPOSE OF PHYSICAL FITNESS TRAINING

Physical fitness training in the Marine Corps has one purpose: to prepare Marines to physically withstand the rigors of combat. All other goals of physical fitness training are subordinate to and must support attainment of this goal. The idea that only infantry or reconnaissance units and their attachments normally face physically demanding combat is wrong. This error must not influence the priority commanders of combat support, combat service support, aviation, and headquarters units give to physical fitness for combat. Physical fitness for combat has a high priority for all Marines. A sound, effective unit program, requiring limited time and material, offers a greater payoff in combat than many more expensive and time-consuming training programs. This handbook describes the concept and provides the ingredients of a program for commanders to use in carrying out one of their most serious responsibilities: ensuring that their Marines are physically ready for combat.

1102. SCOPE OF THE MANUAL

This handbook provides guidance for all leaders, trainers, and planners of physical training programs. It describes unit physical fitness for combat training in the following chapters:

a. Physical Readiness Leadership. Chapter 1 provides guidance to leaders in the conduct of physical combat readiness training.

b. Physical Readiness Training Programs. Chapter 2 discusses how to structure programs to meet physical readiness goals in a variety of combat training situations.

c. Physical Conditioning Activities. Chapter 3 discusses the primary physical conditioning activities which commanders may use to attain readiness for combat goals. Sections in this chapter address foot marches under load, strength-building activities, and activities which build confidence and the aggressive spirit needed in combat.

d. Combat Water Survival. As an amphibious force, Marines cannot ignore the importance of combat water survival training. Chapter 4 describes a program for commanders to conduct this training.

e. Competitive Conditioning Activities. Chapter 5 describes competitive activities which stress the relationship to combat actions, featuring team-building types of competition.

f. Evaluation of Performance During Training. Chapter 6 describes tests which measure individual and unit physical fitness for combat.

g. The Human Body. Chapter 7 covers the structure and functioning of the body.

1103. COMMANDER'S ROLE

Major General Commandant John A. Lejeune in the 1921 edition of Marine Corps Manual expressed our philosophy of leadership. Among other things, General Lejeune stated that "it will be necessary for officers ... to endeavor to enlist the interest of their men in building up and maintaining their bodies in the finest physical condition." It is significant that the Commandant prefaced these writings in 1921 with the statement that his thoughts were greatly influenced by the experience of World War I. We have never varied as a Corps from this belief that one of a commander's most serious responsibilities is to have Marines physically prepared for combat. Senior commanders and their staffs carry out this role primarily by providing command emphasis, including planning, support, and supervision. Commanders at the battalion/squadron and company/battery level execute a program of physical fitness for combat by close supervision and, most importantly, leadership by example.

1104. HISTORY OF MILITARY PHYSICAL READINESS

Every war has revealed our military physical deficiencies during the initial periods of mobilization. This realization followed the Civil War and has recurred regularly with each national emergency.

a. Training programs in each war were geared to the physical need of the era. Success was dependent upon the amount of time available during training to physically prepare Marines for battle conditions. Frequently, casualties in initial engagements were attributed to the inability of our Marines to physically withstand the rigors of combat over rugged terrain and under unfavorable climatic conditions. With adequate preparation, Marines have always handled the test of battle.

b. During World War II, the first physical conditioning doctrine that could be scientifically justified by testing was introduced. As the war progressed, this program was effective in the physical conditioning of millions of Marines for combat.

c. Postwar periods have traditionally been a time of consolidation. Unfortunately, some leaders considered the conditioning phase of training to be a wartime tool. With such a philosophy prevalent between wars, physical readiness was relegated to secondary importance resulting in a serious lowering of combat effectiveness. The initial commitment of Marines in Korea dramatically displayed this failure to recognize the extreme physical demands of warfare. Call-up of large numbers of reserve Marines and pressing them into the war within weeks after call-up mainly caused the loss in effectiveness.

d. Costly lessons learned from our military experiences over a period of years and the course of several wars led to an increasing interest in the physical conditioning of the individual Marine. No longer can we afford to emphasize physical fitness during wartime and de-emphasize it during peacetime. It is evident that, in spite of increased mechanization and modern weapons, physical readiness retains a vital place in the life of each individual Marine and in every unit within the Marine Corps regardless of mission.

e. Commanders are well aware of the need for rugged and well conditioned Marines. The daily demands of housekeeping, maintenance, support, training, operations, and other time-consuming tasks make it necessary for commanders to set aside time for frequent, regular, and vigorous exercise periods.

1105. PHYSICAL DEMANDS OF COMBAT

There are three primary elements of effective physical fitness for combat: lower body strength and stamina; upper body strength and stamina; and a competitive, combative spirit. Cardiovascular functioning is not treated as a separate element since it is a necessary foundation for the other three. Exercises and activities designed to improve Marines' condition in the three primary elements will foster good cardiovascular and respiratory condition.

a. Lower Body Strength and Stamina. Some common demands that combat places on lower body strength and endurance are marching long distances under load and functioning effectively at the destination; moving quickly and evasively under fire; and carrying wounded Marines to safety.

b. Upper Body Strength and Stamina. Some common demands that combat places on upper body strength and stamina arc rapidly emplacing crew-served weapons; handling large-caliber ammunition for extended periods; climbing walls, cliffs, and other high obstacles; and performing field maintenance on air craft or heavy machinery.

c. Competitive, Combative Spirit. While part of this critical aspect of combat readiness is mental and emotional, robust physical condition and a training program which promotes physical aggressiveness greatly enhance a competitive and combative spirit. Activities which force Marines to overcome natural physical fear, which directly leads to fatigue, are particularly valuable.

1106. PHYSICAL EFFECTS OF COMBAT STRESS

It has been thoroughly documented that the added uncertainty and stress of combat have a major physical effect on Marines as well as the acknowledged psychological effect. In other words, in combat, fear equals fatigue. In training, we cannot easily reproduce this combat stress which reduces the effectiveness of individuals and units. However, we can produce fatigue and psychological doubt in training situations by developing a challenging physical training program which stresses the participants. This compels Marines to exercise their ability to continue to function under difficult and trying conditions. Such activities as long foot marches under load and difficult confidence courses train Marines to overcome their own fear and fatigue. Rugged competitive activities such as martial arts and pugil stick fighting prepare Marines psychologically to overcome an opponent as well as their own fear and fatigue. This sort of physically demanding training builds self-confidence and unit morale. The adage that "the unit that sweats more in peacetime will bleed less in battle" certainly applies to its physical training program. Athletic coaches have long followed this approach by such practices as having teams shoot foul shots after wind sprints in order to strengthen their mental conditioning and toughness. Although pushing Marines to their limit is necessary, commanders often set lower standards initially than ultimately desired. The program should then call for steadily increasing the difficulty. A good guide for a starting point is a standard that will be physically and psychologically demanding for no less than three-fourths of the unit. Expressed another way, no more than one-fourth of the unit should be working at less than their limit. The program should progress quickly, augmented, if needed, by remedial training for those unable to meet the unit standards, until the training is challenging for all but the most exceptional Marines in the command. Commanders must never forget that Marines expect to be challenged and thrive on being pushed to their limit.

1107. LIMITS OF PHYSICAL READINESS

A well-conditioned Marine unit has a significant advantage in combat, but this does not mean that physical conditioning can substitute for technical knowledge or good planning. For instance, units must, as discussed in chapter 3, pursue a continuing and challenging program of marches under loads. This docs not disprove the axiom that fighting men, if required to carry over one-third of body weight, will generally become ineffective in battle. Leaders and planners must decide what is essential to be carried and must use an effective logistic distribution system, such as unit trains, to make remaining items available when they are needed. In a similar vein, the valid requirement to foster a competitive and combative spirit in Marines does not excuse leaders from the responsibility to practice tactics which will not needlessly risk lives. Physical readiness for combat is a vital goal, but it is only one part of a unit's overall readiness for the lest of combat.

Section II. Fundamentals of Physical Fitness

1201. FITNESS FOR MARINES

Total combat readiness includes technical proficiency and mental, emotional, and physical fitness. If any one of these attributes is lacking, combat effectiveness suffers proportionately. Without technical fitness, Marines lack the knowledge and skill to fight; without mental and emotional fitness, they lack the incentive and will to fight; and without physical fitness, they lack the physical ability and confidence to fight. Physical fitness in a Marine means a whole and healthy body, the capacity for skillful and sustained performance, the ability to recover from exertion rapidly, the desire to complete a designated task, and the confidence to face any eventuality.

1202. COMPONENTS OF PHYSICAL FITNESS

A sound body, free of disease and defect, does not in itself constitute physical fitness. Before an untrained Marine can be considered physically fit for combat, the following traits must be developed:

a. Strength. Every Marine needs enough strength to perform the heaviest task encountered in routine and emergency activities. The basic areas where strength is required are in the arm and shoulder girdle, abdomen, back, and legs. Muscles increase in size, strength, and firmness with regular and strenuous exercises. Without work, they grow flabby and weak.

b. Endurance. Each Marine needs sustaining power to maintain the maximum ability without undue fatigue. There are two types of endurance:

(1) Muscular Endurance. The Marine needs muscular endurance to fight the enemy under the most tiring combat conditions. Muscular endurance is the ability to perform continuous work over long periods of time. Endurance depends on the bloodstream's ability to deliver large amounts of oxygen and nutrition to the muscle masses and then carry away the waste products quickly.

(2) Cardiovascular-Respiratory Endurance. The development of cardiovascular-respiratory endurance ("wind") is necessary to maintain muscular endurance. Cardiovascular-respiratory endurance depends on the efficiency of the blood vessels, lungs, and heart. The maximum effort a Marine can exert over a period of time is limited by the capacity to absorb oxygen and expel carbon dioxide. The average Marine's cardiovascular-respiratory capacity can be greatly increased by exercise.

c. Agility. A Marine must be able to change direction quickly and as faultlessly as possible. The ability to react instantly and to maintain orientation during rapid changes of body position is important to survival. This important ability may be developed by conditioning exercises that require varied and rapid changes of body position on the ground and in the air, such as obstacle courses.

d. Coordination. Coordination is the ability to move all parts of the body in a smooth, efficient, and concerted effort (commonly called timing). A well-coordinated individual does not make useless movements. An individual who moves with precision and accuracy saves energy. Coordination is best developed by practicing diversified muscular activities and skills affecting all body parts.

1203. TYPES OF EXERCISE

Basically, there are two forms of exercise: isotonic and isometric. Both forms are contained in the activities found in this manual.

a. Isotonic. Isotonic exercises are those in which the expenditure of energy is regulated and released during concentrated efforts. The regulated expenditure of energy is controlled by both the mode of exercise and the individual's effort. This type of effort is common to most exercise and sports. To develop endurance, coordination, and flexibility, isotonic exercise must be used. Strength can also be developed or increased through this type of exercise.

b. Isometric. Isometric exercises arc those in which maximum effort is applied and held until the engaged muscle energy is depleted during a single contraction effort. The individual exerts full force against an immovable object for a relatively short period of time (6 to 10 seconds) and then repeats the contraction several times with a short period of rest between each contraction. This type of exercise develops only strength; therefore, it has limited application.

1204. PRINCIPLES OF PHYSICAL CONDITIONING

Men and women vary in their physical makeup. Physiological (body) function and reaction also vary in proportion to heavy demands placed upon the body. To attain the maximum benefit without sacrificing Marines' welfare, fitness must grow with a careful program of conditioning. To allow for adjustments in body functioning as the conditioning program progresses and to ensure attainment of objectives, the following principles of physical conditioning must be applied:

a. Overload. As strength and endurance increase, the physical load must be increased until the desired level of fitness is reached.

b. Progression. In beginning stages, the load must be moderate. Gradual progression from this low state of fitness to a higher state is possible through application of a progressive program.

c. Balance. An effective program utilizes various types of activities and provides for the concurrent development of strength, endurance, and coordination as well as basic physical skills.

d. Variety. Some programs fail because the routine becomes boring. The most successful programs always include conditioning activities, competitive events, and military physical skill development.

e. Regularity. There is no easy or occasional way to develop physical fitness. Regularity of exercise is a must, with daily exercise preferred.

1205. THREE STAGES OF PHYSICAL CONDITIONING

Unconditioned or poorly conditioned Marines pass through the following stages in reaching the desired state of physical condition.

a. Toughening Stage. This stage is approximately 2 weeks long and is usually characterized by muscular stiffness and soreness followed by recovery.

b. Slow Improvement Stage. This stage is approximately 6 to 10 weeks long and is characterized by slow and steady improvement until the desired level, or a high level, of fitness is attained.

c. Sustaining Stage. This stage goes on indefinitely in order to maintain the level of conditioning achieved by passage through the previous stages.

1206. EFFECTS OF CLIMATIC CONDITIONS

Temperature, both atmospheric and body, affects the physical performance of Marines. Proper maintenance of body temperature through warm up exercise, proper dress in cold weather, and removal or adjustment of clothing in hot weather is necessary for effective performance and health. Climatic factors to be considered are the following:

a. Exercise In High Temperatures and High Humidity. Marines can endure strenuous physical activity in extremely hot temperatures if they are given an opportunity to become acclimated and if they take enough salt and water. It is essential to continue physical training programs in hot climates. Marines can better withstand high temperatures when they are well-conditioned. High humidity combined with high temperatures presents a serious danger. These conditions prevent the natural cooling of the body by the evaporation of perspiration. Training schedules should conform to the provisions of the base commander in respect to wet bulb conditions. Those who conduct training under these conditions should monitor weight loss and be careful to make fluid replacement adjustments.

b. Exercise at High Altitudes. Certain problems are encountered in conditioning Marines stationed in high altitudes because the heart undergoes greater exertion during exercise. It is particularly important that only light exercise be given initially at such altitudes. A Marine's body gradually adjusts to high altitudes within a few weeks. After this adjustment, progressively greater amounts of exercise are possible.

c. Exercise in Arctic Regions. Military duty in the arctic is so arduous that a high level of physical conditioning is essential. Because of the difficulties of carrying on physical conditioning exercises in extreme cold, Marines should be conditioned to the highest level possible before they arrive. A sustaining program will then maintain that level. When exercising in cold weather, Marines should be required to remove excess clothing to prevent them from becoming damp with perspiration.

1207. WARMING UP AND COOLING OFF

It is a fundamental physiological principle that Marines should warm up gradually before taking strenuous exercise. Such action speeds up the circulation to prepare the body to take an overload and helps to prevent injury to muscles and joints after exercising. Marines should be kept mildly active, walking, stretching, or performing some other mild muscular activity until breathing and temperature have returned to normal. Marines should never be allowed to cool off too rapidly; in cool or cold weather, they should put on additional clothing during the cooling-off period.

1208. PHYSICAL ACTIVITY AS AGE INCREASES

Combat makes severe physical demands on Marines. All Marines, regardless of age, must be physically ready to meet these demands. There is no physiological reason to cease exercise or exertion with age. Increased age usually brings increased responsibility which, in many instances, leads to a routine that can become almost devoid of physical activity. The key to fitness with increased age is to continue exercising at a reasonable level and to include exercise of a vigorous type in the daily routine. Older persons who have not regularly maintained a reasonable state of physical fitness require more time than younger persons to become fit. Such individuals usually require a longer period of time to recover from physical effort than younger Marines. If general health is good, evidence shows that older Marines can develop and maintain a degree of fitness which permits vigorous activity and proper performance of duties. It is both the individual's and the commander's responsibility that all Marines continue a daily sustaining exercise program.

Section III. Goals of Physical Readiness Training

1301. NECESSITY FOR PHYSICAL READINESS TRAINING

A very important objective of training is attainment and maintenance of operational readiness. Marines must be physically ready for operations at any time and under all conditions of climate and environment. A combination of training to develop proficiency in physical skills and conditioning to improve strength and endurance results in physical readiness for combat. The degree of physical fitness required of Marines can be acquired through physical exercises in a normal well-coordinated and closely monitored training routine. The performance of purely military duties, such as drills and marching, is not enough to build all the desired areas of fitness. Few recruits are physically fit for the arduous duties ahead of them. The softening influences of our mechanized civilization and the difficulties of conditioning Marines make physical fitness more important than ever before. If Marines are to be fully developed to and maintained at the desired standard of physical fitness, a well-conceived plan of physical readiness training must be a basic part of every training program. Marines cannot be adequately prepared in any other way for the hard work and arduous demands required on the battlefield.

1302. OBJECTIVE OF PHYSICAL READINESS TRAINING

The overall objective of the physical readiness training program is to develop individuals and units who are physically able and ready to perform their duty assignments or missions during training and in combat. To attain the objective of physical readiness, exercise activities must develop the following areas:

a. Strength and Endurance. Developing adequate strength to perform required duties and adequate endurance to sustain activity over a long period of time.

b. Muscle Tone. Developing muscle tone adequate to maintain proper posture and reasonable weight control.

c. Skills. Developing military physical skills which are essential to personal safety and effective combat performance. As skill is developed, agility and coordination are attained. The essential skills are-

(1) Marching Under Load. Marching with individual and unit weapons and equipment.

(2) Running. Distance and sprint running on roads and cross-country.

(3) Jumping. Broad jumping and vertical jumping downward from a height.

(4) Dodging. Changing body direction rapidly while running.

(5) Climbing and Traversing. Vertical climbing of ropes, poles, walls, and cargo nets. Traversing horizontal objects such as ropes, pipes, and ladders.

(6) Crawling. High crawling and low crawling for speed and stealth.

(7) Throwing. Propelling objects, such as grenades, for a distance with accuracy.

(8) Vaulting. Surmounting low objects, such as fences and barriers, by use of hand assists.

(9) Carrying. Carrying objects and employing man-carries.

(10) Balancing. Maintaining proper body balance on narrow walkways and at heights above normal.

(11) Falling. Contacting the ground from standing, running, and jumping postures.

(12) Surviving in Water (Or Other Specialized Situations) Using water survival techniques.

d. Character Traits. Jnstilling character traits which help accomplish military missions to include-

(1) Confidence. Confidence develops through achieving progressively more difficult tasks as physical ability grows.

(2) Aggressiveness. Participation in combative activities and contests develops desire and willingness to overcome an opponent.

(3) Reaction Under Pressure. Competitive contests and game situations train Marines to think and to act quickly while under pressure.

(4) Teamwork. Teamwork develops through competitive events in which a number of Marines must coordinate their efforts to accomplish a physical task.

1303. BENEFITS OF EXERCISE

The benefits of exercise are not always understood. Some of the more important results of exercise are listed below:

a. Improved Muscle Tone. Muscular tone improves and, at the same time, muscular strength and endurance are built up.

b. Cardiovascular-respiratory Endurance. Cardiovascular-respiratory endurance, or wind, improves through a process of opening up dormant lung capacity to absorb greater amounts of oxygen.

c. Circulation. Circulation of the blood speeds up and extends to a greater portion of the body as exercise forces the blood to service all parts of the body. The efficiency and effectiveness of the heart, lungs, and blood vessels improve.

d. Flexibility. A wider range of muscular movement is possible and rapidity in physical skills grows.

e. Elimination of Body Waste. Bending and twisting the body and the general speedup of body processes caused by exercise regulate and help eliminate body wastes.

f. Tension. Working off excess nervous energy and relief from daily worries and cares relieve tension. Participation in exercises leaves little time for worry.

g. Sleep. Sleep improves because muscles are healthfully tired after about of exercise. A by-product of sound sleep is relief of tension.

h. Obesity Control.. Control of obesity (fat) is made possible by using up excessive amounts of fat-producing food elements.

i. Injury Susceptibility. Susceptibility to injury is reduced through exercise. Muscles, tendons, and joints are strengthened. Injuries such as hernia, back strain, and joint sprains are less likely to occur if muscles are maintained in proper tone.

Section IV. Leadership Roles

1401. PSYCHOLOGICAL LEADERSHIP

The full development of a Marine's resources is not all physical. To be effective in developing physical readiness, leaders must realize that mind and attitude are also important to success. The more important psychological considerations are to—

a. Promote Understanding of the Value of Physical Readiness. A desire to be physically ready should be created in all Marines. Motivation is increased and Marines take greater interest in their individual physical fitness if they understand the value and benefits of vigorous exercise. When Marines realize their efforts are an investment in their own personal welfare, it should not be difficult to obtain their cooperation. Marines should understand the objectives, the benefits, and the value of each type of exercise activity in their program. They should also understand the relation of physical readiness to survival in combat.

b. Maintain a Positive Approach. Physical readiness training for combat is strenuous and demanding. It is a responsibility of leadership to create an atmosphere where all desire to participate fully. This attitude should be fostered. A negative approach must not be identified with physical readiness training, even with those having difficulty. Only in unusual cases should fear of punishment be the motivating factor behind good performance. For those few who cannot keep up or attempt to malinger, an effective remedial program is essential.

c. Seek Cooperation and Develop Morale. In a program placing maximum physical stress upon individuals, it is necessary to gain their cooperation. Favorable reaction is enhanced by proper planning and organization, challenging requirements, use of competition, and application of a progressive program resulting in physical fitness. As physical fitness grows, morale also grows.

1402. COMMAND AND SUPERVISORY FUNCTIONS

a. Command Functions. Commanders should take the following actions to support physical readiness training:

(1) Lead by personal example.

(2) Instill command interest and indicate to subordinate personnel the importance of this training to the welfare of the organization.

(3) Allot sufficient time for the achievement of objectives and monitor the use of such allotted time. The substitution of other training or routine duties for scheduled physical readiness training is unsound and unwise.

(4) Assign and properly utilize qualified personnel to supervise and conduct physical readiness training. If leaders are not competent, take action to ensure they become competent quickly. Reassign those who do not meet this standard to other duties.

(5) Make necessary facilities and funds available to support a program to develop physical readiness within all personnel.

(6) Measure the physical fitness of individuals and units in order to evaluate progress and to determine if the program is successful.

b. Supervisory Functions. Leaders responsible for planning, conducting, and supervising physical readiness training should take the following actions:

(1) Prepare physical readiness training schedules which apply the principles of physical conditioning and which aim for a particular type of program plan.

(2) Provide for wide participation of as many Marines as possible. All Marines, regardless of position or age, will benefit from regular exercise. In some instances, special efforts are necessary to overcome obstacles to regular and frequent training. Special effort is also necessary to ensure remedial conditioning. Such conditioning should occur for those who are physically substandard and after extended absence due to leave, sickness, injury, and travel.

(3) Prevent waste or unwise use of time allotted for physical readiness training. Time-wasters include unprepared instructors; assignment of one instructor to a group larger than a platoon; progression which does not keep pace with the physical development of the Marines; extreme formality; inadequate equipment or facilities which require waiting turns to exercise; and lengthy rest periods between exercises which interfere with the application of overload.

(4) Ensure that the program contains vigorous physical activity. Such activity places progressively greater demands upon the body during each exercise session and also over the duration of the training program. To be of benefit, exercise must tire the muscles and cause the heart to increase its rate of beat.

(5) Set an overall objective for each physical fitness program.Observe the training as necessary to ensure that the established objectives are being achieved.

(6) Observe physical readiness training to insure the use of a positive approach. To implement a positive attitude, small-unit leaders and instructors should personally set the example; have an understanding, fair, and sympathetic attitude; recognize individual differences; and motivate Marines toward their best effort.

(7) Guide and inform small-unit leaders and instructors concerning approved techniques, directives, and literature. As necessary, arrange for local training of instructors to include clinics, conferences, schools, and demonstrations.

(8) Determine the effectiveness of physical readiness training by personal participation in and observation of training, analysis of field inspection reports, and analysis of individual physical fitness test scores. Scores may be combined to reflect the fitness of the unit.

1403. SMALL-UNIT LEADERS AND INSTRUCTORS

a. Responsibility. The instruction and conduct of physical readiness activities are the function of company/battery commanders, platoonleaders and persons assigned as instructors. Experience has proved the effectiveness of physical fitness development when conducted in company-and platoon-size units under direct control of the leader with overall supervision by the parent-unit commander. For example, all the platoons of a company may exercise at the same time under the general supervision of the company commander, with each platoon conducting the assignment separately and under its own leadership.

b. Leader's Assignment. A small-unit leader or an instructor in a school or training activity is assigned to a combat unit or to a support unit. In this assignment, the leader is responsible for all training to include physical readiness training. In a different situation, the noncommissioned officer (NCO) or officer may beassigned as a full-time physical readiness instructor. This contrasts to a unit leader assignment where only part of the time is devoted to such training. In either case, the leader will hold an important and vital position for the physical fitness of Marines.

c. Leader's Training. Leaders may come to the assignment either fully or partially trained. It may be their first responsibility for the development of physical fitness. If they have had previous training through experience, make certain that their information is supplemented with study of this manual. If they have had professional training in physical education during civilian life, but no military experience, they should also use this manual to learn the methods used by the Marine Corps. A new leader should take advantage of various ways to learn including attendance at leader training courses, self-study, practice, and discussion with more experienced leaders.

d.Leader's Objective. As a physical readiness training instructor, the leader has two general objectives. The first is to motivate Marines to want to be physically fit. The second is to conduct a program that will develop a high degree of physical fitnss. Motivated Marines will react enthusiastically to such a program. It aids greatly in achieving local program objectives.

e. Leader's Personal Fitnes. A unit leader who must instruct and demonstrate physical activities must be in physical condition to do the job without undue physical stress.The leader should be able to do those things that must be demonstrated. The leader's strength, endurance, posture, and skill should set the example. This does not mean that the leader must excel, as other Marines do not expect championship performance. However, they do expect, and deserve, a creditable showing of fitness for the job.

f. Leader's Knowledge. The leader must have three types of knowledge to properly administer physical readiness training. Theyare- -

(1) Knowledge of Marines. The leader must understand Marines, know how to lead and motivate them, understand how they learn, and apply this knowledge wisely in the day-to-day training situation.

(2) Understanding of Body Functioning. A more intelligent exercise program results from understanding and applying the principles which govern physical conditioning of the body. The leader with such knowledge can better prescribe, adjust, and regulate exercise types, amounts, and progression to attain fitness.

(3) Understanding Exercise Activities. The leader needs to understand the contribution each type of physical activity makes to physical fitness, and how to use each activity to develop fitness. Skill to demonstrate and lead the various activities is a necessary part of technique and is invaluable to the instructor or small-unit leader.

Chapter 2
PHYSICAL READINESS TRAINING PROGRAMS
Section I. Development of a Program

2101.GUIDANCE FOR PLANNERS

This chapter instructs planners on physical readiness training procedures. It contains program planning guidance including factors to consider when developing programs; steps in assembling a program; definitions of activity packages and systems of exercise; selection of activity packages; and selection of systems used in implementing those packages.

2102. MARINE CORPS MISSION

The mission of the Marine Corps is to seize and defend advanced naval bases and to perform other missions as the President of the United States may direct. Traditionally, to perform this mission, Marines have been projected into the area of operations by Navy vessels and have attacked hostile beaches over the shore in landing craft, assault amphibian vehicles, and helicopters. Recently, performance of this task has been somewhat complicated by the introduction of the concepts of maritime prepositioning and the air-landed Marine expeditionary-brigade. Under these concepts, Marines, landed at airfields in foreign countries, will join their equipment at commercial ports. The distance between these airfields and ports is often 10 or more miles. In time of international crisis, it can be expected that transportation assets will not be adequate. Marines from all elements of a Marine Air-Ground Task Force may have to march from the airfield to the site of their equipment. All Marines must be physically conditioned to rigors of conducting foot marches with individual weapons and equipment.

2103. EVALUATION OF FITNESS

The Marine Corps mission is such that all Marines must constantly and consistently achieve a high level of fitness that prepares them for the demands of combat. While Marines have many different MOS's, all Marines must be prepared for the demands of marching under load and performing basic infantry tasks, such as rear security and patrolling. While the physical fitness test remains our universal measure of individual fitness, it is the commander's responsibility to observe and evaluate the unit's ability to perform effectively in combat. Often the commander's most effective evaluation tool is weekly physical training sessions. The commander should participate in these sessions and see that they are fully integrated into the unit physical conditioning program.

2104. INTEGRATION OF TRAINING

Training time must be used efficiently and wisely. Every opportunity to integrate physical training into other training activities should be seized. For example, foot marches under load can be integrated into the normal daily activities as a way to move from one training or work site to another. This integration reflects realism in training and should be used to maintain overall proficiency. To accomplish this requires imagination and enthusiasm on the part of the commander. The area of operations in which the training is conducted determines what can be integrated and how. If possible, training should also consider the terrain and climate of the area in which the unit will subsequently conduct training or operations. Training should include familiarization with special equipment and the application of specialized techniques to tactical principles. Activities such as gun drills and command post emplacement drills also have physical conditioning value, particularly if supervised and done competitively or measured against time standards.

2105. ACTIVITIES AT THE UNIT LEVEL

Physical readiness training is a command responsibility and is generally conducted in Marine units at the company and platoon levels. Battalion/squadron commanders must constantly supervise, making sure that all Marines are physically prepared for combat.

a. Physical readiness training should usually be conducted weekly by each battalion/squadron-level unit, using events designed for all personnel (e.g., marches under load, battalion/squadron runs, military field days, etc.).These events enable the commanders to demonstrate personal leadership, and observe and evaluate the physical condition and combat physical readiness of their Marines. This exercise period should be designed to elevate the unit's morale and emphasize unit identity through the wearing of uniforms and the carrying of battalion and company guidons. The battalion/squadron commander should make every effort to be a visible participant in the unit's physical readiness program on a weekly basis. These activities could be conditioning marches under load or a weekly calisthenics/unit run.The suggested times for these activities are early on Monday morning or late on Friday afternoon after the training week has been completed. The uniform may be physical training gear orutilities and boots.

b. Commanders should be cautious of long runs in boots and utilities. This is not to suggest that units should not run in utilities, but to remind commanders that boots are designed for marching, not running. If commanders choose boots and utilities, they should select off-road routes which will provide a cushioning effect.

c. Units should be made well aware of the commander's intent before the unit falls out for exercise. If the exercise period is conducted early in the morning, it need not necessitate an extensive shower/clean-up period afterward, so long as all Marines from the commander on down remain dirty. Commanders should inform their Marines not to fall out for these sessions in clean, pressed uniforms.

d. The physical readiness training of headquarters units is often difficult to manage. Often physical training is left for individuals or sections to develop and manage on their own. This approach can be effective, but if not supervised, it can result in a "paper" program. Headquarters unit commanders should endeavor to conduct weekly integrated training which will enable them to evaluate the physical readiness of their personnel. Often this training has to be scheduled before or after normal working hours to maximize participation of the entire unit.

2106. STEPS IN PLANNING

To implement workable and effective programs (as directed by CMC ALMAR 261/87) the planner must-

a. Determine the Type of Program Needed. Marine units are inherently different in organization and mission. The physical readiness program must be tailored to the mission and to the current physical condition of most unit personnel. Programs to meet this need are of the following types:

(1) **Developmental Programs.** Marines in a beginning or poor state of physical readiness need a program which will develop strength, endurance, physical skills, and character traits which are beneficial to successful accomplishment of military missions. Such programs should be applied progressively to rise gradually to a peak of fitness and skill.

(2) Maintenance Programs. Once Marines reach the sustaining stage of conditioning, their goal is then to maintain this level by participation in a maintenance program.

(3) Remedial Programs. The term "remedial" is usually applied to those individuals or groups who possess substandard physical fitness. For example, a remedial physical conditioning program could be applied to persons who are overweight, who fail to reach physical fitness test standards, or who have missed extended periods of conditioning due to illness, injury, extended hospitalization, or other absence. Saturday training sessions are designed as remedial sessions for all individuals who either miss the daily training or have failed to meet adequate standards. Note that personnel in light duty or no-duty status should be expectedto attend all training sessions, observing, supervising, or assisting as needed.

b. Determine the Time Required. The amount of time for training operations varies considerably. However, every unit can find time to conduct physical readiness training. Frequent (i.e.,daily) physical training of short duration-30 to 60 minutes-is preferred over occasional longer periods. Other demands for training time are so urgent that every minute of time allotted for physical readinesstraining should be used. Determine the time required per week anddivide it into daily blocks.

c. Organize for Various Group Size. It is essential to stress exercise rather than formality. Marines must complete the program where they are- -on the training field, in the motor pool, on the range, next to the classroom, in the office area, in the shop, aboard ship, or elsewhere. Although desirable, it is not always possible to assemble company-size units for physical training. Platoon-size groups are appropriate for the proper conduct of physical conditioning activities. Certain situations may require exercise programs for section- or squad-size units. This manual outlines programs for all situations and types of organizations.

d. Allow for Weather and Exercise Area. In programing and scheduling, the climate and terrain often govern the selection of activities.

(1) Weather changes cause differences in temperature, rainfall, wind chill, and snow. These changes should be anticipated as they dictate the type of program. Alternate plans should be part of the schedule. Nevertheless, normal weather changes and rain should not drive training inside.
(2) Local terrain and available exercise areas may also influence the selection of activities and the type of program which it is possible to support. Some activities can be completed in nearly any area.
e. Plan for Seasonal Change. As most physical readiness training is conducted outdoors, it is necessary to recognize seasonal change. A program should be divided into fall, winter, spring, and summer parts. Seasonal change also causes change in light. For example, an early morning program started in the summer will have ideal lightconditions, yet in the fall or winter, darkness will occur at that same hour and interfere with the conduct of the program. Develop programs in seasonal blocks, and make provisions for anticipated changes.

f. Consider Needed Facilities. An excellent program can be conducted with practically no facilities since there are exercises which require no equipment. However, a better program can be developed when supported by certain facilities and items of equipment. Proper command support, plus ingenuity, will solve this problem. (Items of equipment, when necessary to support the recommended exercises, are included in the chapters on exercise activities.)

g. Specify Appropriate Uniform. The uniform worn for exercising depends upon the season of the year, the state of the weather, and local regulations. Whenever possible, Marines should be dressed alike. Undershirts are preferred as the upper garment when the weather permits. A uniform that restricts the free movement of the body should not be worn when exercising.

h. Consider Availability of Instructors. Leaders who can lead and direct the scheduled activity must be available. Organizational units should train junior officers and noncommissioned officers down to squad or section leaders to instruct and lead the various activities.

i. Select Activity and System. With the type of program needed and the objectives in mind, select an activity package or an integrated training package for each day's scheduled physical training. At this same time, the system to be used in employment of the selected activity or activities must be determined. This selection affects equipment, areas, instructors, transportation, and other support requirements.

j. Secure Command Participation and Support. Prepare and brief the commander to assure full understanding of the objectives and administration of the program. The full participation and support of the commander will greatly improve the level of success.

k. Supervise Execution. Determine needs, publish the program, and supervise its execution as a necessary part of developing the unit's training schedule.

2107. SELECTION OF ACTIVITIES AND SYSTEMS

a. Exercise Activities. The planner must determine exercise activities which will be appropriate to include in the program. An exercise activity is a single means of exercise usually identified by the name applied; for example, running, log exercises, and obstacle course. Many exercise activi ties can become part of activity packages.

b. Activity Packages. Many of the physical activities described in this manual are arranged in prescribed sequences and are known as activity packages. An activity package is a number of exercises of the same type, assembled as a group or a set, and arranged in a specific sequence. Exercise packages are organized in such manner that not more than 15 minutes will be required to complete the execution of any package. Each type of activity is explained in laterchapters. The number of available drills, tables, or circuits; the manner of organization; and the contribution each makes to the total program are covered. Full understanding of this information will greatly assist in developing effective programs. Various designations are used to identify exercise packages; for example, conditioning exercises when arranged in a set order are known as drills, and other packaged activities are designated as tables or circuits. The following are activity packages:

(1) Activity packages for groups:

o Conditioning Drill Two
o Conditioning Drill Three
o Rifle Drill
o Log Drill
o Grass Drill
o Running Tables
o Guerrilla Tables
o Circuit-Interval Table
o Combatives Tables
o Relay Tables

(2) Strength Circuits:

o Fixed Circuit
o Movable Circuit
o Simplified Circuit (Circuit Interval Table)

(3) Activity packages for individuals:

o The 6-12 Plan
o Weight Training
o Isometric Exercise

c. Advantages of Using Packages. These of exercise packages simplifies scheduling and conducting of exercise and results in the following benefits:

(1) Schedule development is simplified as the planner assembles packages which will satisfy the training objective. There is no need to deal with selection of individual activities or to be concerned about the amount of time to be expended on each.

(2) Any 15-minutc period, and in some cases less time, can be scheduled or used to perform an activity.

(3) If longer periods of time are available or if the objective demands, several packages can be assembled to provide a more complete period of activity.

(4) Marines are assured a balanced set of exercises or activities as each package is carefully arranged to reach all muscle groups.

(5) The instructor can concentrate on the conduct of a vigorous workout as the type and duration of the activity have already been determined.

d. Nonpackaged Activities. Several types of activity are not packaged. Activities in this category are conditioning marches, unit runs, obstacle courses, combat water survival swimming, teamcontests, and team sports. These activities can be scheduled in combination with packaged activities, or they may be scheduled separately. Often they require a longer period of time. Most nonpackaged activities require a 50-minute period to satisfactorily complete their objective. For conditioning marches, a few hours are required. The benefits of these activities should not be overlooked as some desirable objectives cannot be attained without them.

e. Systems of Exercise. Several methods or systems of organizing exercise and activity packages can be used. Each system is based upon a specific organization as follows:

(1) Single Activity System. The unit leader immediately assumes command of the unit at the beginning of the exercise period. The leader moves the unit to a predesignated exercise site at double time, forms the unit in a circle around the leader, groundsclothing and equipment as appropriate, and quickly moves into the exercise routine. There is usually no time to teach; therefore,the Marines must know the activity to be used. At the conclusion of a 5- to 15-minute period, the unit leader returns the unit to the instructional area at double time and releases the unit for the next scheduled activity.

(2) Progressive Activity System. All Marines (company or platoon) complete activities in the same order during the period. For example, Drill One is followed by dual combatives, and finally a 1-mile run. This system is usually progressive from a warm up activity, such as Drill One, to an activity which contributes in a major way to one of the objectives, such as aggressiveness development through combatives or cardiovascular development, such as running.

(3) Rotating Activity System. Set up the same number of activities or stations as there are platoons in the company. Each platoon rotates through each station in turn. With four platoons in a 50-minute period, about 10 minutes can be devoted to each station. With three platoons, approximately 15 minutes may be spent at each station. Activities must be of a type that can be covered in the time allotted. For example, with three platoons, Station 1 could be Conditioning Drill One; Station 2, running; and Station 3, a team contest.

(4) Circuit System. Set up a number of stations to provide various types of exercise equipment and/or items of apparatus. The idea is to keep all Marines busy and exercising vigorously for a short period of time at each station. The fixed strength circuits and the movable strength circuits are examples of this system. Station changes must be rapid, and the exercise must be started quickly after each change. Since the objective is to exercise at top speed, the motivation comes from frequent change of activity by moving to another station. Rotation by station groups continues until all Marines have covered all stations.

(5) Interval System. This system stresses the development of strength and endurance. It involves heavy work for a given distance within a specified time, alternated with lighter work and recovery, but never stopping during the workout. This procedure is repeated, and the intensity is increased gradually as physical condition improves, but always with adequate recovery. The important concept is stress, recover, stress, recover, and so on. This system is often applied through running, but other activities of a continuous nature may also be used. An example is the Circuit-Interval Table.

f. Steps in Selecting Activities and Systems. The planner should follow these steps.
(1) Evaluate the needs of one unit with emphasis upon unit mission, objective, and time available.
(2) For each day, select an activity package or a combination of activities which will contribute to the objective.
(3) Then determine the system to be used in implementing the selected activities. There are several possible choices and much opportunity for flexibility in program development. For example, the weekly program may contain various exercise packages, systems, and time periods. Additional guidance is contained in following chapters.

2108. SAMPLE WEEKLY PROGRAMS

The variety of weekly programs given shows available alternatives. Programs and schedules other than those illustrated may be assembled to provide schedules to fit any situation. Many other scheduling combinations are possible through use of the packaged activities, varied time periods, and different types of assigned duty.
a. Sample A. Figure 2-1 illustrates a 60-minute time allotment in which the single activity system is used with activity packages that vary from day to day. In this program, 5 hours of training are scheduled for the week. This program needs some type of integrated unit training to supplement it each week. A bimonthly conditioning march of 10 miles in 3 hours is suggested to round out this program. The Saturday session is designed for remedial physical training.
NOTE: A physical training activity must be conducted at a minimum of once every 72 hours. After 72 hours, muscles begin to atrophy.

Figure 2-1 is available in the illustration section at the end of the book

b. Sample B. A more comprehensive weekly program (fig. 2-2) illustrates the scheduling of various-length periods, a variety of activities, and the use of three systems during the week. In this schedule, 3.5 hours are included for the week. The following should be noted concerning this schedule:

Figure 2-2 is available in the illustration section at the end of the book

(1) On Monday, a 60-minute period is available for physical readiness training. Three 20-minute packages are scheduled. Each platoon, under its own leadership, will progress through each scheduled activity in turn.
(2) On Tuesday, Wednesday, and Thursday, only 30-minute periods are available. Here each platoon leader supervises each platoon in a single activity system.
(3) On Friday, the company commander directs training to exhibit personal leadership example and supervision. In this example, only one strength circuit and one cross-country course are available, and each will accommodate only one platoon. For this reason, Conditioning Drill One station is included. A platoon rotates to each station within the 60-minute period.

(4) The Saturday session once again is designed for remedial physical training.
(5) Bimonthly, a 10-mile/3-hour march under load will be necessary to round out this program.
NOTE: Cardiovascular/respiratory exercise to improve endurance must be of long enough duration (20 minutes or more) to induce and maintain oxygen debt.

c. Sample C. A third sample schedule (fig. 2-3) illustrates the use of the single activity system for 5.25 hours of training. This schedule includes sustaining-type activities for a unit that has passed through both the toughening and slow improvement stages of conditioning. The use of competitive activities is featured to holdinterest and provide self-motivation. A bimonthly 10-mile/3-hour march under load will be necessary to round out this program. The Saturday session once again is designed forremedial physical training.

Figure 2-3 is available in the illustration section at the end of the book

NOTE: The body will respond to hard work or physical training without soreness and stiffness provided the work or physical training is conducted regularly.
NOTE: Exercises of short duration increase strength and bulk but do little to effect endurance.
NOTE: The times shown in the charts are participation times.

Section II. Remedial Physical Training

2201. NEED FOR ATTENTION

Remedial physical training is a process by which physically substandard individuals are conditioned to meet standard levels of performance. To achieve a full degree of operational readiness, it is necessary to bring all Marines up to the prescribed standard. Experience has demonstrated that some Marines have difficulty due to a poor state of fitness, obesity, or lack of motivation. Attention to these individual deficiencies will help improve unit combat readiness.

2202. IDENTIFICATION OF INDIVIDUALS

The company or battery commander identifies persons who cannot achieve the prescribed level of proficiency through the physical fitness test or as a result of observation during training. The commander notes particular weakness of body parts as indicated by failure of certain test events. These Marines are then placed in a special remedial program either at platoon, company/battery, or battalion/squadron level. Extra time is devoted to overcoming the weakness. These extra periods of conditioning may be during or after duty hours, as determined by the commanders. Other persons need reconditioning as result of hospitalization. Restoring physical fitness to damaged areas of the body is accomplished through progressively graded physical activities under professional supervision, not by the physical fitness trainers.

a. Types of Deficiencies. Physical deficiencies which can be corrected by exercise fall into several categories.

(1) Lack of Strength in One or More Body Parts. The major muscle areas concerned are the arms and shoulder girdle, back, abdomen, and legs.

(2) Lack of Overall Endurance. Usually, there is deficient muscular and cardiovascular endurance.

(3) Deficiency in Coordination and Agility. In these cases, physical skill is not developed to a satisfactory degree in activities such as crawling, running, jumping, climbing, traversing, vaulting, pushing, pulling, lifting, and carrying.

(4) Overweight or Underweight. Either condition may interfere with physical fitness and mission accomplishment. Lack of exercise is not always the cause. The cause may be malfunction of normal physiological functions or it may be poor health habits such as over- or under-eating, lack of adequate rest, or overconsumption of alcohol.

(5) Lack of Motivation. Not all persons are motivated to attain or maintain **a** desirable state of fitness. Some Marines do not understand the importance of physical fitness, some find proper exercise too difficult, and others find it inconvenient.

b. Causes of Deficiencies. There are several causes for Marines being deficient in physical fitness:

(1) Absence of exercise.

(2) Exercise which fails to develop all muscle groups and components of fitness.

(3) Exercise which is not vigorous enough or which lacks progression.
(4) Injury or illness which depletes fitness.
(5) Inadequate amounts of sleep or rest.

c. Methods of Detection. The commander has several means by which to detect physical deficiencies:
(1)
from those who pass. Make an analysis to determine the cause of failure as revealed by test scores.
(2) Observe Marines as they perform physical tasks-
o Marines who have difficulty during training or in physical types of work,
o Marines who have become obese and therefore experience difficulty.
(3) Be alert to those Marines who are often on sick call, returnees from hospitalization, or newly assigned.

2203. ADMINISTRATION OF REMEDIAL ACTION

a. Group. Attitude Marines who are singled out as being physically deficient are self-conscious and are not always convinced they need extra help. Within a deficient group, motivation may be low. These Marines must be convinced that a remedial program which is tailored to their needs will help them and will eliminate their deficiencies. Encouragement is often needed and desirable.

b. Leadership. The leader of this group must study individuals and know their deficiencies. The unit leader should counsel Marines individually, observe them closely as they progress through the remedial program, maintain records, and adjust the program as required.

c. Measurement.. Whenever improvement in performance is noted, Marines should be measured by the physical fitness test either individually or by group. When testing reveals an individual to be satisfactory, that individual should be released from the remedial group. There may be exceptions to this policy in the case of Marines who are overweight or in the execution of an exercise program prescribed for some special purpose.

d. Organization of a Remedial Group
(1) A remedial group is usually a small group of Marines within a company or battalion. In some situations, it may be a larger group numbering 50 to 75 Marines. In the case of a larger group, divide the Marines initially into subgroups according to ability. Prescribe exercise loads commensurate with their ability. General conditioning is sufficient in the early part of the program to qualify Marines who are on the borderline. Qualification will reduce the size of the group, permitting more individual and specialized attention to those who remain.
(2) As the program progresses, regroup individuals who have like deficiencies in order that they can concentrate on their weakness. For example—

(a) One group may be weak in the arms and shoulders as revealed by failure in the chin up event. This group, with an assistant instruction charge, can work on pull-ups, rope climb, pushups, rifle or logexercises, horizontal ladder, and similar types of exercise.

(b) Another group may be weak in general endurance as revealed by thc3-milc run. This group could profit from participation in conditioning drills, running, grass drill, and strength circuit.

(3) It may be necessary to form some groups to overcome weaknesses in skills, such as an inability to throw, to quickly change direction while running, to crawl rapidly, or to carry a load. Lack of coordination or lack of practice may cause these deficiencies. In this instance, instructors must provide an opportunity to practice and correct poor form and other errors as they are noted.

Chapter 3
PHYSICAL CONDITIONING ACTIVITIES
Section I. Marching Under Load

3101. GENERAL

Few physical fitness activities are as directly related to readiness for combat as foot marches under load. In addition to the obvious physical conditioning and unit cohesiveness benefits, marches under load prepare most Marines for numerous foreseeable tasks in combat. The idea that only infantry and reconnaissance units actually require training marches to prepare them for combat is clearly a mistake. In addition to the very real possibility of Marines from all types of units being used to fill combat-depleted infantry units, it is likely in maritime prepositioning force operations that aviation, combat service support, and command element personnel will have to march with weapons and equipment, from the arrival airfield to the equipment marshalling site. In American coastal cities, the harbor is nearly always 10 or more miles from the airport. lt is difficult to imagine any different condition in a lesser developed country .A further combat-related benefit is that foot marching under load is the most effective way to develop leg strength in the context of a unit training program. In his classic study of troop performance in World War II, Men Against Fire, S.L.A. Marshall noted that the greatest single weakness of replacements of all occupational fields was leg strength. Truck drivers must push trucks which are stuck in mud, and headquarters communicators must climb hills to emplace antennas. The demands on the lower body in combat apply to all Marines, and foot marching is one activity which can help build these muscles while being conveniently integrated with the rest of the training program.

3102.TRAINING GOALS

The standard for success of a foot march is very simple to measure: did the unit arrive at the destination at the prescribed time with Marines in condition and required equipment present to accomplish the mission?A progressive program can increase a unit's readiness. lt can instill pride by increasing distances and rates of march and by selecting routes over increasingly challenging terrain as the Marines become better conditioned. However, the ability to execute the mission at the conclusion of the march must remain the standard of success.

3103.MOVEMENT PLANNING

The basic considerations in planning a foot march are the mission, tactical situation, terrain and weather, and the units to participate. The success of the march will depend largely upon the thoroughness with which it is planned. A successful march is characterized by adherence to prescribed routes and time schedules, the efficient employment of the means available, and the ability of the unit to accomplish its assigned mission upon arrival at the destination.

a. Movement Orders. Movement planning culminates in the preparation and issuance of an operation order prepared in the standard, five-paragraph format. Necessary annexes are attached to furnish detailed information required for the movement. Written movement orders are rarely prepared at company level. A discussion of the annexes is contained in this chapter.

b.March Planning. March planning, as discussed here, is the planning conducted at battalion level. March planning may be organized into the following steps:

(1) **Preparation and Issuance of the Warning Order.** In order to afford subordinate units the maximum possible time to prepare for a pending move, a warning order containing all available information about the march is issued. The amount of planning time available will determine the time of issuance and the content of the warning order.

(2) Estimate of the Situation. In his estimate, the commander considers the mission, terrain, weather, time, and space factors, available routes, available transportation for the movement of equipment and/or shuttling of Marines, enemy capabilities, disposition of own forces, physical condition and training of Marines, and courses of action available to the command.

(3) Organization and Dispatch of a Reconnaissance. Party very march plan is based on as thorough and complete a ground reconnaissance as time and the situation will permit. Map and aerial reconnaissance are valuable in formulating a plan, but are not a substitute for ground reconnaissance. Route reconnaissance is accomplished by a reconnaissance party which usually consists of a reconnaissance element, an engineer element from the attached or supporting engineer unit, and a traffic control element. Unit standing operating procedures generally establish the basic composition of the reconnaissance party. lt is modified as necessary to meet the requirements of a particular march. A recommended method of reporting information obtained by the route reconnaissance party is shown in figure 3-l.The minimum information required from the reconnaissance party is—

(a) Available routes and conditions.(Routes may be specified by higher headquarters.)

(b) Recommended rate of march.

(c) Selection of start point and release point, or confirmation of the suitability of start point and release point previously selected by map reconnaissance.

(d) Confirmation of location of the assembly or bivouac area.

(e) Location of critical points on the route.

(f) Distance between critical points on the route and total distance from start point to release point.

Figure 3-1 is available in the illustration section at the end of the book

(g) Location of obstacles and estimation of necessary Marines and equip- ment needed to repair and maintain routes.

(h) Number of guides required and their location on the route.

(4) Development of Detailed Movement Plan. Some of the elements included in a detailed movement plan arc-

(a) Organizing the Column. To facilitate control and scheduling, units will be organized into serials and march units and given an order of march. In determining the order of march, the march planner must consider the enemy situation and the desirable order of arrival of the units at the destination. Where dispersion is required, a unit may be organized into two or more columns, each assigned a different route.

(b) Using Reconnaissance Information. Results of the route reconnaissance will be used to select the route(s); determine the start point, critical points along the route(s), and the release point; and select the rate of march.

(c) Determining March Computations. March unit pass time is based on the strength, formation, and rate of march. The pass time of the marching columns, plus necessary time distance computations, will be used to determine the completion time of the march.

(d) Drafting of Road Movement Table. Using the completed march computations, a draft road movement tabic is compiled.

(e) Checking the Plan. Using the draft road movement table and a road movement graph, the movement plan is checked to ensure that it conforms to the directive of the higher headquarters and the battalion commander's instructions.

(5) Preparation and Issuance of the Road Movement Order. After the plan has been checked and approved by the commander, an operation order is prepared and issued. The operation order may be in written form or issued orally, and is accompanied by a road movement table, overlay and/or strip map, and appropriate administrative details.

(6) Road Movement Table. A road movement table, prepared as an annex to the operation order, provides serial commanders with arrival and clearance times at critical points along the route of march. It also provides the column commander with information as to the proposed location of elements of the column at various times.

(7) Overlay/Strip Map. An overlay serves the normal purpose and should show, as a minimum, the present location of units, route of march, critical points, and the new location of units at the destination. A strip map is a schematic diagram of the route of march and shows landmarks and critical points with the distances between them. A strip map may be issued as an annex to the road movement order, in addition to or in lieu of an overlay.

c. March Computations. Prior to issuance of the operation order, the S-3 must verify time and space computations as they provide him with the necessary data for the preparation of a road movement table.

(1) Time-Distance. Time distance (TD) is determined by dividing the distance to be traveled (D) by the rate of inarch (R):

TD = D/RWhere TD = hours D = distance in kilometers R = kilometers per hour

(2) Length of ColumnThe length of column (LC) is used to determine the pass time of a column. The sum total of the following two parts determines the length of column. (See pars, (a) and (b)).The two parts are the space occupied by the Marines alone (including the distance between Marines) and the sum of the distances between the units of the column (column gap).

(a) The length of column of Marines alone is determined by multiplying the number of Marines by the appropriate factor selected from the table below. The length of column docs not include distances between units. LC Marines = No. of Marines x factor.

(b) The total distance in meters between units is determined as follows:

Step One. Determine the number of gaps between serials (total serials minus one).

Step Two. Multiply the number of serial gaps from Step One by the length (in meters) between respective units.

Step Three. Determine the number of gaps between march units (total march units minus one, minus the number of serial distances).

Step Four. Multiply the number of march unit gaps from Step Three by the length (in meters) between respective units.

Step Five. Add the totals from Step Two and Step Four to get the total meters for the column gap.

EXAMPLE:A battalion foot column is organized into 12 platoon-sized march units and 3 company-sized serials. REQUIRED: total column gap distances when there are 100 meters between serials and 50 meters between march units.

(3) Pass-Time.Pass-time (PT) is the time a unit takes to pass a specified point. For foot columns, the pass-time is determined by applying the following formula:PT (minutes) = LC x FACTOR (for appropriate rate of march).

PASS-TIME FACTORS - FOOT TROOPS

.0150 for 4.0 km/h
.0187 for 3.2 km/h
.0250 for 2.4 km/h
.0375 for 1.6 km/h

EXAMPLE: Determine the pass-time of a unit whose length of column is 1,500meters and is marching at a rate of 4 km/h.PT (min) = 1,500 x .0150 (the factor for 4.0 km/h) = 22.5 mm.

(4) Completion Time. Completion time is the time of day that a march will be completed. Completion time is determined by using the following formula: Completion Time = SP time + TD + PT + Scheduled Halts (other than normal hourly halts).

EXAMPLE: A column's starting point (SP) time is 0700.The time-distance is6 hours and 40 minutes. Pass-time of the column is 30 minutes. A 35-minutelunch halt has been scheduled. What is the completion time of the march?Employing the 24-hour clock system, the formula can be applied as follows for simplified addition of the times:

HrMin SP time.................0700
TD......................0640
PT......................0030
Lunch halt...............0035

Completion time.........13 105or 1445

The march will be completed at1445.

(5) Experience Tables. Based on previous movements made by a unit, data is accumulated to facilitate march planning. Such data includes approximate pass-times lor various elements of the battalion. The S-3 can utilize these data rather than computing them each time a march is scheduled. Such experience tables serve to reduce the time required to complete the computation phase of march planning. Matter appropriate to the unit standing operating procedure should be integrated therein.

d. Road Movement Graph

(1) A road movement graph is a time-distance diagram used in planning, preparing, or checking road movement tables, and for controlling marches. The graph provides a visual representation of a march plan so that conflicts and discrepancies may be prevented in the planning stage, before congestion occurs on the route. It is not usually issued as a part of the order. Road movement graphs may be applied to small units, to movements of a single column, or to a large organization scheduling separate elements, moving by various means, with different rates of march, over one or more routes.

(2) To construct a road movement graph, use the following steps: (See fig. 3-2.)

Figure 3-2 is available in the illustration section at the end of the book

Step One. Determine the number of hours available for the march. Designate the lower left corner of a sheet of graph paper as the SP time or the earliest even hour before the march is to begin. Select a scale and plot the hours available in sequence from the left on the horizontal axis.

Step Two. Determine the distance to be moved in kilometers. Indicate the SP at the lower left corner of the graph sheet and, using an appropriate scale, plot the number of kilometers on the vertical scale from the SP to the release point. Indicate the location of critical points on the vertical scale.

Step Three. At the proper distance from the SP, draw a horizontal line indicating the location of the RP. Indicate by a vertical line the hour when the movement must be completed. Plot lines representing route restriction, if any, at the proper distances and times on the graph.

Step Four. Determine the pass-times of foot and motor elements of the column.

Step Five. Starting at the SP at the specified hour, plot the movement of the head of the leading element. If the rate of march is 3 kilometers per hour, the lead element will have moved 1.5 kilometers in 30 minutes, 3 kilometers in 1 hour, etc. Plot the trace of the lead man to the RP. The last Marine will cross the SP in pass-times after the lead man. Measure this time on the graph and plot the trace of the last Marine of the column. The lines describing the head and tail of a march serial are parallel. Indicate the time subsequent serials reach the SP and plot the trace of the head and tail. Check to see that the plan complies with all restrictions and orders. If there are any violations or conflicts, the plan may be made to conform by changing the order of march, the starting time of the serials, the planned rate of march, or the organization of the column.

e. Conversion to the Metric System. The adoption of the metric system for expressing linear distances requires a simple means of conversion.

(1) To accomplish this, use the following tables:

o Multiply inches x 2.54 to obtain centimeters. Multiply centimeters x .39 to obtain inches,

o Multiply yards x .91 to obtain meters. Multiply meters x 1.1 to obtain yards,

o Multiply miles x 1.6 (or 8/5) to obtain kilometers. Multiply kilometers x .62 (or 5/8) to obtain miles.

(2) Linear distances can be accurately measured on maps which have been converted to the metric system by using a straightedge graduated in centimeters (cm).Thc centimeter scale can be used in lieu of the graphic scale when the scale of the map is shown, as follows:

Map Scale:	One cm Equals:
1:500,000	5,000 meters
1:250,000	2,500 meters
1:100,000	1,000 meters
1:50,000	500 meters
1:25,000	250 meters

3104. INDIVIDUAL LOAD

Since the backbone of Marine ground combat forces is the infantry, tactical mobility is largely dependent on the marching ability of the individual Marine. The load that each Marine must carry is the greatest impediment to mobility. In general, the commander must make every effort to reduce the individual load of his Marines to the absolute minimum. While the load has the greatest effect on Marines marching on foot, the value of carrying the minimum essential supplies and equipment applies for all Marine forces.

a. Fighting Load. The fighting load consists of items of clothing, equipment, weapons, and ammunition that are carried by, and essential to, the effectiveness of the combat Marine and the accomplishment of the immediate mission. The fighting load should not weigh more than 40 pounds. The commander must ensure that his Marines' loads- -and his own- -be stripped to the bare minimum. The addition of small, light-weight items in excess of the limit should not be tolerated, since cumulatively, these items will take a toll in energy. Every extra pound a Marine must carry decreases his combat effectiveness.

b. Existence Load. The existence load consists of items other than those in the fighting load that are required to sustain or protect the combat Marine, that may be necessary for increased personal and environmental protection, and that are not normally carried by the individual. The existence load is normally carried by the Marine's unit. Marching, but not engaged in combat, a Marine may be expected to carry a maximum of 50 pounds of supplies and equipment.

c. Considerations

(1) The primary consideration is not how much a Marine can carry, but how much he can carry without impaired combat effectiveness-moral or physical. The combat strength of a unit is not counted simply in numbers of Marines, but in the numbers of willing and physically able Marines.

(2) It is generally better to risk temporary inconvenience from lack of health and comfort items than to exhaust Marines due to overloading.

(3) It is a fundamental truth that men become physically exhausted more quickly when under the stress of combat. Marines must be required to carry less into battle than they are conditioned to carry in training. Marines should be conditioned for carrying weight, but should be equipped in combat for fleetness of foot.

(4) A common mistake is to base the fighting load on thc gcar and supplies necessary to meet every contingency. The commander cannot reasonably expect the Marines to carry enough gear for every possible eventuality. The items to be carried must be based on reasonable expectations.

(5) It is the commander's responsibility to procure the transport to carry any additional gear. As a rule of thumb, a rifle company, or unit of similar size, requires one 5-ton truck and trailer in support to carry additional gear. In cold weather, or other conditions in which the necessary personal gear increases, this requirement will increase.

(6) The commander must ensure that the supply system provides, on a dependable and timely basis, the balance of essential supplies and equipment not carried by the unit. Marines must feel confident that they will be supported with the necessary supplies and equipment.

(7) In training, commanders must implant in their Marines the pride in operating under austere conditions. To effectively demonstrate the need for this spartan mentality, the commander must personally set the standard. Marines should be trained in field craft, foraging techniques, and the use of caches and field expedients. Maximum use should be made of captured stores.

3105. MARCH TECHNIQUES AND PROCEDURES

a. Organization. A command executing a march is organized into march units, march serials, and march columns, as necessary for control.

(1) March Units. A march unit is a unit of command which moves and halts at the command of a single commander. The march unit normally corresponds to one of the smaller Marine units such as a platoon or company.

(2) March Serials. A march serial consists of one or more march units organized under the senior officer and given a numerical or alphabetical designation to facilitate control. A serial is usually a battalion, but may be a company if the battalion is marching alone.

(3) March Columns. A march column is composed of elements of a command moving over the same route. It may be composed of one or more serials. To facilitate control, a column commander is designated. A column is normally a regiment or larger unit, but may be a battalion if marching alone.

b. Order of March. The order of march is determined by the mission, terrain, and the commander's desired order of commitment of units into action.

c. Control and Coordination Measures. The commander establishes initial control of the march by designating control measures in his march order. The most frequently used measures include:

o Start point and release point.
o Time at which head or tail of column passes the start point.
o Rate of march.
o Order of march.
o Route of march.
o Location of command post.
o March communications.

d. The Pacesetter. The pace-setter is an experienced individual carrying the same load as the majority of the Marines and marching from 4 to 10 meters at the head of the column. (See fig. 3-3.)The pace-setter's primary duty is to maintain the rate of march ordered by the column commander.

(l) Setting the Pace. The pacesetter does this by establishing the pace (length of step) and cadence (steps per minute) to obtain the prescribed rate of march. He should be of medium height so normal strides will be taken. Overstriding or understriding tends to tire leg muscles quickly and affects the combat efficiency of marching Marines. The officer marching at the head of the column supervises the pacesetter to ensure that he takes normal strides and maintains a uniform cadence.

Figure 3-3 is available in the illustration section at the end of the book

(2) Reducing Accordion Effect. Accordion effect occurs when the column alternately lengthens and shortens, causing the rear elements of the column to double time to maintain distance. It is caused by changes in the rate of march by the lead elements, usually after passing a slowing obstacle. To reduce this effect as much as possible, lead elements should slow their rate of march for a sufficient time after crossing obstacles to allow the rear elements to maintain the prescribed distance without running. The rear elements can signal the head when the last man has cleared the obstacle and the rate can be resumed. Since some accordion effect is nearly inevitable, the order of march should be periodically rotated to prevent units in the rear from becoming physically exhausted at the conclusion of the march.

e. Length, Rate and Pace

(1) **Length of March.** The length of march varies depending on terrain and weather, enemy situation, and physical and mental condition of Marines. The normal length of march for a 24-hour period is from 20 to 32 kilometers (12 to 20 miles) marching from 5 to 8 hours at a rate of 4 kilometers (2.5 miles) per hour (km/hmi/h).A march in excess of 32 kilometers in a 24-hour period is considered a forced march. Well-trained units that have been progressively trained will be able to achieve a unit goal of 60 miles in a 3-day period.

(2) Rate of March. The same factors of terrain, weather, load to be carried, and condition of Marines affect the rate of march. The commander considers these and selects a rate which will place his unit at the destination in the shortest possible time in condition to accomplish the mission. Unit standing operating procedure should state the standard rate of march for that unit over normal terrain for both road and cross-country movement and in both day and night. The commander then modifies this rate if the situation requires. Normally prescribed standards are as follows:

Figure is available in the illustration section at the end of the book

(3) Pace and Cadence. The normal pace is 30 inches. A pace of 30 inches and a cadence of 106 steps per minute result in a speed of 4.8 kilometers per hour or 3 miles per hour and a rate of 4 kilometers per hour or 2.5 miles per hour if a 10-minute rest halt per hour is taken. Since the pace of each man may vary, the cadence may have to be adjusted to maintain the prescribed rate. Slope of ground and surface footing affect pace. A normal cadence is relatively simple to maintain on level or gently rolling terrain on firm footing. Mud, sand, loose gravel, and hills all greatly affect pace.

f. Halts. Halts during a day's march are taken at regular intervals to rest personnel and adjust loads .Halts are specified by standing operating procedures or march order. Such factors as midday heat or enemy action may force the adoption of long midday halts or night marches. At long halts, each unit disperses to a previously selected location near the route of march.

(1) Normal Time and Frequency. Under normal conditions, a 15-minute halt is made after the first 45 minutes of marching. Following the first halt, a 10-minute halt is made after each 50 minutes of marching. Variations of this schedule are made when a scheduled halt time occurs when passing through a built-up area or when cover and concealment are required by the tactical situation and none is available. Observation posts may be established, if required for security of the unit during halts.

(2) Actions at Halts. All units in the column should be halted at the same time. At the halt signal, Marines should move to the side of the road, staying within the immediate vicinity of their unit. Marines should remove or loosen gear and sit or lie with feet elevated. Commanders inspect Marines and equipment, and corpsmen administer medical treatment, as required.

g. Communications. The four primary means of communications used in controlling foot marches are radio, visual, sound, and messenger.

(1) Radio. Radios are normally used for communications between platoons and higher headquarters in the march column. Radio transmissions should be held to the absolute minimum necessary for control and should be at minimum power required.

(2) Visual.. Visual communications may include flashlights, luminous markers, panel sets, pyrotechnics, and hand and arm signals. When contact with the enemy is not expected, the loss of light discipline from some of these signals may be a lesser security risk than the risk of disclosure from radio transmissions.

(3) Sound. Sound communications include voice and such signaling devices as horns and whistles. Some of their uses include assembling Marines at the conclusions of halts and providing warnings of air or chemical attacks.

(4) Messengers. Messengers are particularly effective during periods of radio silence and reduced visibility. Messages should be simple and brief to preclude mistakes by the messenger.

h. Security. Both active and passive measures are used against attack by enemy aircraft and indirect fire weapons during movement. Active measures include the use of air sentries and organic and attached weapons in accordance with the unit air defense standing operating procedure. Passive measures include the use of concealed routes and assembly areas, night marches, and extended interval between elements of the column. With imagination and planning, all of these measures can be practiced in training without detracting from the physical conditioning benefit of the march. In fact, such drills as rapidly taking cover during simulated attacks and then quickly resuming the march can prove physically challenging.

i. Reconnaissance. Reconnaissance determines in advance critical points along the march route such as bridges, fords, and obstacles in order that congestion or delay may be held to a minimum and local security provided. Paragraph 3103.C describes organization of the reconnaissance party.

3106.TACTICAL MARCHES

These marches are movements of Marines and equipment not in direct ground contact with the enemy but expecting early ground contact either en route or upon arrival at the objective. Some characteristics of this type of movement include security elements to the front, flanks, and, if necessary, to the rear of the march column. Additional considerations include the selection of concealed areas on defendable terrain for start and release points and rest stops. In training, infantry units of regimental size or smaller will normally derive the greatest benefit from tactical marches. The security requirements usually dictate a slower rate of march and, therefore, reduce the physical conditioning value for other types of units which would seldom move in combat with the expectation of early ground contact with the enemy. Two key points should be remembered regarding tactical marches in a training program.

a. Security. Although security requirements often slow the rate of march, it is a valid and important training objective for infantry units to work toward reducing march times while still practicing sound tactical security. March standing operating procedures, good land navigation skills, and simple but complete march orders contribute as much to this goal as physical conditioning. The idea that proceeding slowly enhances security is seldom true. Keeping the enemy off balance by rapid movement frequently is the best guarantee of security.

b. Training. In peacetime training, it is critical that support personnel who normally are located with infantry units in combat, participate completely with infantry units in forced march training. Not only must they not slow the infantry rate of march because of inferior conditioning, but they must be thoroughly familiar with the infantry unit's standing operating procedure for security on the march. Personnel from within the ground combat element such as artillery forward observer teams and combat engineer units, can be easily identified for this training and should be habitually associated with the same infantry units. Personnel from other elements, such as low-altitude air defense teams and radio battalion detachments, are more difficult to identify and earmark for support of the same infantry unit. Their normal mode of employment is in vehicles and their units frequently provide general support vice direct support. This does not, however, lessen their requirement to be prepared to provide support to dismounted infantry over terrain or in a tactical situation not conducive to vehicles. Such training needs to take place, and planning and cooperation among the elements of the MAGTF can overcome the obstacles.

3107.ADMINISTRATIVE MARCHES

These marches are movements of Marines and equipment made when no enemy interference is expected except from aircraft or indirect fire. During these marches, units are administratively grouped for ease of control and speed of movement.The term administrative foot marches should not be taken to imply that tactical considerations are disregarded.There are simply less stringent security measures used than when ground contact with the enemy is a distinct possibility.The normal formation is the route column with one file moving on either side of the road and with negligible security. Air sentries are used and dispersion between individuals and units is practiced. Although roads usually provide the fastest route, there may be reasons to conduct these marches cross-country (particularly as part of a training program). Administrative marches are the type of foot marches most practical for training programs for non-infantry units. These marches may also play a key role in the physical conditioning program of infantry units, because of the greater speeds possible due to lessened security requirements.

3108. FORCED MARCHES

A forced march is a march which covers a greater distance than 32 kilometer (20 miles) in a 24-hour period. Normally the extra required distance should be achieved by increasing the number of hours marched in a day beyond the norm (8 hours) rather than increasing the rate of march beyond the norm (4 kilometers per hour/2.5 miles per hour). However, there will be occasions when the situation or mission demands an increase in the rate of march.

a. Maximum Recommended Distances. The maximum recommended distances for forced marches are-56 kilometers (35 miles) in 24 hours; 96 kilometers (60 miles) in 48 hours; or 128 kilometers (80 miles) in 72 hours.

b. Sample Time Breakdown. A sample time breakdown for a forced march of 52 kilometers, beginning at first light is as follows:

Figure is available in the illustration section at the end of the book

3109. NIGHT MARCHES

Night marches are characterized by closed formations, more difficult control and reconnaissance, and a slower rate of march than day marches.

a. Control. Control is increased by reducing the distance between individuals and units, and by using connecting files to maintain contact between platoons and companies. Connecting files are normally constituted from personnel from the rear most march unit.

b. Safety on Roads. Because of reduced visibility, night marches on roads used by vehicles require attention to safety procedures to prevent accidents. Commanders should not think these safety measures apply only to peacetime training. In combat, nearly all vehicles operate with blackout lights at night, making visibility even more difficult than in garrison driving. If consistent with the tactical situation, the following measures can reduce the possibility of marchers being struck by vehicles:

(1) Use off-road trails and routes as much as possible.

(2) Guards to the front and rear of march units should be marked with strips of luminous tape and carry red-filtered flashlights. These measures are generally consistent with light discipline, while still affording reasonable warning to approaching drivers.

Section II. Conditioning Drills (One, Two, and Three)

3201. GENERAL

a. Description and Objective. Conditioning drills are calisthenic exercises. Each drill contains seven exercises organized and numbered in a set pattern. Each drill takes 15 minutes to complete. The objective of conditioning drills is to exercise all major muscle areas in order to develop strength, endurance, coordination, and flexibility.

b. Area and Equipment. Any level area is satisfactory for conduct of the drills. Drills One and Two contain ground exercises. If ground conditions are unsatisfactory, Drill Three should be used as it contains no ground positions. Usually, no equipment is required;however, if the group exceeds a platoon in size, an instructor's stand is necessary.

c. Formation. The extended rectangular formation is used in this drill. (See app. A.)

d. Starting Level and Progression. The starting level is six repetitions of each exercise. An increase of one repetition for each three periods of exercise in which the drill is performed is an acceptable rate of progression. This rate is continued until 12 repetitions can be completed. To maintain, continue the drill at 12 repetitions. To progress, move to 6 repetitions at a more difficult drill which exercises the same muscle group. Progression can also be gained by moving from Drill One to Drill Two, as Drill Two is more demanding.

e. Starting Positions. Starting positions vary with the exercise and are explained in each exercise.

f. Leadership. A principal instructor demonstrates and leads the drills. The instructor must be familiar with leadership techniques peculiar to conditioning drills to include the exercises, commands, counting cadence, cumulative count, formation, method of teaching the exercises, and utilization of assistant leaders.

g. Use With Other Programs. Conditioning Drills One, Two, and Three reach all major muscles of the body. They are easy to learn and to perform, and they are simple to administer and supervise. These features, coupled with the short time required for completion, the fact that no equipment is necessary, and adaptability to most areas of execution, make these drills possible in any programs.

3202. CONDITIONING DRILL ONE

This conditioning drill is similar to the calisthenics drill commonly known throughout the Marine Corps as the "Daily Seven."The "Daily Seven" may be used as a substitute for Conditioning Drill One.

a. Exercise l:High Jumper

(1) Starting Position. Feet separated shoulder width, knees flexed, body bent forward at the waist, arms aligned with the trunk and hips, elbows locked, palms facing, fingers extended and joined, head and eyes to the front.(See fig. 3-4, A.)(Elbows remain locked throughout the exercise.)

Figure 3-4 is available in the illustration section at the end of the book.

(2) Cadence. Moderate.

(3) Movement. A four-count exercise: at the count of—

(a) ONE -- Take a slight jump into the air, swinging the arms forward and up to shoulder level.

(b) TWO -- Take a slight jump into the air and swing the arms downward and back, returning to the starting position.

(c) THREE -- Take a vigorous leap into the air, swinging the arms forward and up to an overhead position, momentarily looking skyward, on returning to the ground the knees are flexed, head and eyes return to the front.

(d) FOUR--Repeat the action of count two.

b. Exercise 2: Bend and Reach

(1) Starting Position. Feet spread more than shoulder width, arms overhead, elbows locked, palms facing, fingers extended and joined, head and eyes to the front.(See fig. 3-4, B.)

(2) Cadence. Moderate.

(3) Movement. A four-count exercise: at the count of—

(a) ONE-Bend at the knees and waist, swing the arms straight downward and reach between the legs. Touch the ground as far to the rear as possible and look to the rear. (Elbows remain locked throughout the exercise).

(b) TWO-- Recover sharply to the starting position.

(c) THREE--Repeat the action of count ONE.

(d) FOUR-- Repeat the action of count TWO.

c. Exercise 3: Pushup

(1) Starting Position. Front leaning rest position. To assume this position there is a silent one-two count:on the silent count of one, assume the squatting position, heels together, elbows locked inside the knees, hands flat on the ground directly beneath the shoulders; on the silent count of two, thrust the legs to the rear, toes and heels together, body straight from head to heels.(Seefig. 3-4, C.)

(2) Cadence. Moderate.

(3) Movement. A four-count exercise: at the count of—

(a) ONE -- Flex the elbows lowering the body until the thick portion of chest touches the ground.

(b) TWO -- Raise the body until elbows are straight and locked.

(c) THREE-- Repeat the action of count ONE.

(d) FOUR -- Repeat the action of count TWO. (On returning to position of attention, the silent one-two count is used in reverse).

d. Exercise 4: Trunk Twister

(1) Starting Position. Feet are spread more than shoulder width apart, fingers placed behind neck, thumbs pointing downward, elbows back. (See fig. 3-4, D.)(Elbows remain well back throughout the exercise).

(2) Cadence. Slow.

(3) Movement. A four-count exercise: at the count of-

(a) ONE-- Keeping the knees locked and the back straight, bend forward at the waist sharply, with a slight recovery.

(b) TWO-- Twist the trunk to the left vigorously at the waist, keeping the elbow back. The left elbow is higher than the right.

(c) THREE-- Twist vigorously to the right, so the left elbow comes under the right.

(d) FOUR-- Straighten sharply to the starting position. NOTE: Do not attempt to touch the elbows to the knees on counts two and three.

e. Exercise 5: Squat Bender

(1) Starting Position. Feet arc spread less than shoulder width apart, hands on hips, thumbs in small of back, elbows back. (See fig. 3-4, E.)

(2) Cadence. Moderate.

(3) Movement. A four-count exercise: at the count of—

(a) ONE-- Assume the squatting position, maintain balance on the balls of the feet, with trunk erect thrust arms forward to shoulder level, elbows locked, palms down.

(b) TWO-- Recover to starting position. Elbows are well back.

(c) THREE-- Keeping the knees locked, bend forward at the waist, touching the ground in front of the toes.

(d) FOUR-- Vigorously recover to the starting position.

f. Exercise 6: Body Twist

(1) Starting Position. On the back, arms extended sideward on the ground, palms down. The legs are raised to a near vertical position, feet together, knees locked.(See fig. 3-4, F.)

(2) Cadence. Slow-fast.

(3) Movement. A four-count exercise: at the count of-

(a) ONE-- Lower legs slowly to the left until they touch the ground near the left hand, keeping the knees straight and shoulders on the ground.

(b) TWO-- Recover the starting position by quickly raising the legs, keep knees straight and feet together.

(c) THREE-- Repeat movement of count ONE, except the movement is to the right side.

(d) FOUR-- Recover sharply to the starting position.

g. Exercise 7: Stationary Run

(1) Starting Position. Position of attention. (See fig, 3-4, G.)

(2) Cadence. Fast.

(3) Movement.

(a) At the command of execution, start running in place at double-time cadence, lifting the left foot first time cadence. Follow the instructor as he counts two repetitions of cadence;

e.g., 1, 2, 3, 4--1, 2, 3, 4. The instructor then gives informal commands such as FOLLOW ME, running on the toes and balls of the feet, keeping the back straight, speeding up the cadence to a sprint, raising the knees high, leaning forward at the waist, and pumping the arms vigorously.

(b) To halt the exercise, the instructor will count two repetitions of cadence as the left foot strikes the ground: 1, 2, 3, 4 - 1 , 2, 3, HALT.

NOTE: When counting cadence, the instructor counts only as the left foot strikes the ground. The duration of the exercise is approximately 1 1/2 minutes.

3203. CONDITIONING DRILL TWO

a. Exercise 1: Jumping Jack

(1) Starting Position. Feet separated more than shoulder width, arms overhead.(See fig.3-5, A.)

(2) Cadence. Moderate.

(3) Movement. A four-count exercise: at the count of~

(a) ONE-- Jump to position with the feet together and assume the squatting position, swinging the arms sideward and downward, placing the hands palms down on the ground, elbows locked inside the knees.

(b) TWO-- Recover to the starting position by jumping to the side straddle and swinging the arms sideward overhead.

(c) THREE-- Repeat the action of count ONE.

(d) FOUR-- Recover to the starting position.

b. Exercise 2: Turn and Bend

(1) Starting Position. Side straddle, arms overhead.(See fig. 3-5, B.)

(2) Cadence. Moderate.

(3) Movement. A four-count exercise: at the count of-

(a) ONE-- Turn the trunk to the left and bend forward over the left thigh, attempting to touch the fingertips to the ground outside the left foot. Keep the left knee straight. On successive repetitions attempt to touch farther and farther to the side.

(b) TWO-- Recover to the starting position.

(c) THREE-- Turn the trunk to the right and bend forward over the right thigh, trying to touch the hands to the ground outside the right foot. Keep the right knee straight.

Figure 3-5 is available in the illustration section at the end of the book.

(d) FOUR-- Recover to the starting position.

c. Exercise 3: Eight-Count Pushup

(1) Starting Position. Position of attention. (See fig. 3-5, C.)

(2) Cadence. Moderate.

(3) Movement. An eight-count exercise: at the count of—

(a) ONE-- Assume the squatting position, palms on the ground directly beneath the shoulders, elbows locked inside the knees.

(b) TWO-- Thrust the legs to the rear assuming the front leaning rest position.

(c) THREE-- Flex the elbows until the thick portion of the chest touches the ground.

(d) FOUR-- Raise the body on a straight plane until the elbows are locked.

(e) FIVE-- Repeat the action of count THREE.

(f) SIX-- Repeat the action of count FOUR.

(g) SEVEN-- Recover to the squatting position as in count ONE (elbows locked inside the knees).

(h) EIGHT—Return sharply to the position of attention.

d. Exercise 4: Turn and Bounce

(1) Starting Position. Feet spread more than shoulder width apart, arms sideward at shoulder level, palms up. (See fig. 3-5, D.)

(2) Cadence. Slow.

(3) Movement. An eight-count exercise: at the count of—

(a) ONE-- Turn sharply to the left as far as possible, then recover slightly.

(b) TWO-- Again turn to the left as far as possible and recover as in ONE.

(c) THREE-- Repeat the action of count TWO.

(d) FOUR-- Recover sharply to the starting position.

(e) FIVE—Turn sharply to the right as far as possible, then recover slightly.

(f) SIX—Again turn to the right as far as possible and recover as in FIVE.

(g) SEVEN-Repeat the action of count SIX.

(h) EIGHT-Return to the starting position.

NOTE: The head and hips remain to the front throughout the exercise and the knees and elbows are locked at all times.

e. Exercise 5: Squat Stretch

(1) Starting Position. Attention. (Sec fig. 3-5, E.)

(2) Cadence. Moderate.

(3) Movement. A four-count exercise: at the count of—

(a) ONE-- Squat, placing the hands on the ground about 12 inches in front of the feet.

(b) TWO— Keeping the fingertips on the ground, straighten the knees completely and raise the hips.

(c) THREE-- Recover to position ONE.

(d) FOUR-- Recover to the starting position.

f. Exercise 6: Leg Circular

(1) Starting Position. On the back, arms stretched sideward, palms down, feet raised foot from ground, knees straight. (See fig 3-5,F.)

(2) Cadence. Slow.

(3) Movement A four-count exercise: at the count of—

(a) ONE-- Swing the legs as far as possible to the left, keeping the knees straight and the legs together.

(b) TWO-- Swing the extended legs over head with the thighs as close as possible to the trunk.

(c) THREE-- Swing the legs as far as possible to the right.

(d) FOUR-- Recover to the starting position.

g. Exercise 7: Stationary Run

(1) Starting Position. Position of attention. (See fig. 3-5, G.)

(2) Cadence. Fast.

(3) Movement

(a) At the command of execution, start running in place at double-time cadence, lifting the left foot first. Follow the instructor as he counts two repetitions of cadence; for example: 1, 2, 3, 4--1, 2, 3, 4. The instructor then gives informal commands such as FOLLOW ME. Running on the toes and balls of the feet, keeping the back straight, speeding up the cadence to a sprint, raising the knees high, leaning forward at the waist, and pumping the arm vigorously.

(b) To halt the exercise, the instructor will count two repetitions of cadence as the foot strikes the ground: 1, 2, 3, 4-1,2, 3, HALT.

NOTE: When counting cadence, the instructor counts only as the left foot strikes the ground. The duration of the exercise is approximately 1 1/2 minutes.

3204. CONDITIONING DRILL THREE

a. Exercise 1: Side Straddle Hop

(1) Starting Position. Position of attention. (See fig. 3-6, A.)

(2) Cadence. Moderate.

(3) Movement. A four-count exercise: at count of-

(a) ONE-- Take a slight jump into the air, moving the legs sideward (more than shoulder width apart); at the same time, swing the arms overhead (to an overhead position) clapping the palms together.

Figure 3-6 is available in the illustration section at the end of the book.

(b) TWO--Take a slight jump into the air, swing the arms sideward and downward returning to the starting position.

(c) THREE-- Repeat the action of count ONE.

(d) FOUR-- Repeat the action of count TWO.

b. Exercise 2: Back Bender

(1) Starting Position. Standing, feet 12 inches apart, fingers placed behind the head. (See fig. 3-6, B.)

(2)Cadence. Slow.

(3) Movement. A four-count exercise: at the count of-

(a) ONE--Bend the upper trunk backward, raising the chest high, pulling the elbows back, and looking upward. Keep the knees straight.

(b) TWO-- Recover to the starting position.

(c) THREE-- Repeat the action of count ONE.

(d) FOUR-- Recover to the starting position.

c. Exercise 3: Squat Thrust

(1) Starting Position. Position of attention. (See fig. 3-6, C.)

(2) Cadence Moderate.

(3) Movement. A four-count exercise: at the count of—

(a) ONE-- Assume the squatting position; heels together, placing the hands flat on the ground, shoulder width apart, elbows locked and inside the knees.

(b) TWO-- Thrust the legs to the rear, assuming the front leaning rest position, body in line from head to toe, heels and toes together.

(c) THREE-- Return to the squatting position as in ONE.

(d) FOUR-- Return to position of attention.

d. Exercise 4: Side Bender

(1) Starting Position. Feet are spread more than shoulder width apart, arms are raised sideward and overhead, thumbs interlocked palms to front, fingers extended and joined, elbows locked. (See fig. 3-6, D.)

(2) Cadence. Slow.

(3) Movement. An eight-count exercise: at the count of—

(a) ONE-- Bend to left as far as possible, then recover slightly.

(b) TWO-- Again bend to the left as far possible, then recover slightly.

(c) THREE-Repeat the action of count TWO.

(d) FOUR-- Recover sharply to the starting position.

(e) FIVE-- Bend to the right as far as possible, then recover slightly.

(f) SIX-- Again bend to the right as far as possible, then recover slightly.

(g) SEVEN-- Repeat the action of count SIX.

(h) EIGHT-- Recover sharply to the starting position.

NOTE:Keep the elbows and knees locked throughout the exercise. The bend should occur to the side and not the front.

e. Exercise 5: Knee Bender

(1) Starting Position. Feet are spread less than shoulder-width apart, hands on hips, thumbs in small of back, elbows back. (See fig 3-6, E.)

(2) Cadence. Moderate.

(3) Movement. A four-count exercise: on the count of—

(a) ONE-Do a knee bend, lean trunk forward at the waist, thrust arms between legs until the extended fingers touch the ground palms to the ground, hands 6 inches apart.

(b) TWO~Recover sharply to the starting position.

(c) THREE-Repeat the action of count ONE.

(d) FOUR-Repeat the action of count TWO.

f. Exercise 6: Bottoms Up

(1) Starting Position. Front leaning rest position. A silent one-two count is used as in the pushups. (See fig. 3-6, F.)

(2) Cadence. Moderate.

(3) Movement A four-count exercise: at the count of-

(a) ONE-- With thc wcight on the hands, and knees locked, jump forward bringing the feet as close to the hands as possible; look to the rear.

(b) TWO-- Keeping the knees locked, thrust the legs backward assuming the front leaning rest position.

(c) THREE-- Repeat the action of count ONE.

(d) FOUR-- Repeat the action of count TWO.

g. Exercise 7: Stationary Run

(1) Starting Position. Position of attention. (Sec fig. 3-6, G.)

(2) Cadence. Fast.

(3) Movement

(a) At the command of execution, start running in place at double time, lifting the left foot first. Follow the instructor as he counts two repetitions of cadence; for example: 1, 2, 3, 4--1, 2,3, 4. The instructor then gives informal commands such as FOLLOW ME, running on the toes and balls of the feet, keeping the back straight, speeding up the cadence to a sprint, raising the knees high, leaning forward at the waist, and pumping the arms vigorously.

(b) To halt the exercise, the instructor will count two repetitions of cadence as the left foot strikes the ground: 1, 2, 3, 4 - 1, 2, 3, HALT.

NOTE: When counting cadence, the instructor counts only as the left foot strikes the ground. The duration of the exercise is approximately 1 1/2 minutes.

Section III. Rifle and Log Drills

3301. RIFLE DRILL

a. Description and Objective. Rifle exercises are conditioning exercises performed with a rifle. Each drill contains six exercises (fig. 3-7) and they are numbered in a set pattern. The drill takes 15 minutes to complete. The objective of rifle drills is to exercise the arms, shoulders, and back muscles in order to develop strength and endurance, particularly in the upper body. In units without rifles, log drills may be substituted.

b. Area and Equipment. Any level area is satisfactory for conducting this drill. Each Marine completes these exercises from a standing position and no ground contact is required. Each Marine will need a rifle and, if the group exceeds a platoon in size, then the instructor will need an instructor's stand.

c. Formation. The extended rectangular formation is used in this drill. (See app. A.)

d. Starting Positions. Starting positions vary with the exercises and are explained in each exercise. As in all set conditioning drills, the command used to start the exercise is STARTING POSITION, MOVE. The following directions apply to rifle drill.

(1) In those exercises which start from the rifle downward position, on the command MOVE, execute port arms as prescribed in NAVMC 2691, Drill and Ceremonies Manual, and then assume the starting position. The command to return to the men to the position of attention at the conclusion of the exercise is POSITION OF ATTENTION, MOVE.

(2) In exercises which terminate in the rifle downward position, on the command MOVE, execute the position followed by order arms as prescribed in NAVMC 2691.

(3) In the exercises which terminate in a position other than the rifle downward position, Marines first assume the rifle downward position before executing port arms and order arms.

(4) These movements are executed without command. This procedure promotes uniformity, but precision is not expected. To be effective, rifle exercises must be strenuous enough to tire the arms, but not to the point where the arms cannot move with precision.

e. Leadership. A principal instructor demonstrates and leads the drill. He must be familiar with leadership techniques for conditioning exercises and the peculiar techniques for rifle drill.

3302. EXERCISES PERFORMED WITH RIFLES

The exercises of rifle drill are outlined in the following paragraphs,

a. Exercise 1: Foreup, Behind Back

(1) Starting Position. Rifle downward, feet together.(See fig. 3-7, A.)

Figure 3-7 is available in the illustration section at the end of the book.

(3) Movement. A four-count exercise: at the count of—

(a) ONE-- Swing the arms forward and upward to the overhead position. Inhale.

(b) **TWO**-- Lowcr thc riflc to thc back of the shoulders. Exhale.

(c) **THREE**-- Recover to position ONE and inhale.

(d) **FOUR**-- Recover to the starting position and exhale.

b. Exercise 2: Lunge Side, Turn and Bend

(1) Starting Position. Rifle downward, feet together.(See fig. 3-7, B.)

(2) Cadence. Moderate.

(3) Movement. An eight-count exercise:at the count of—

(a) ONE-- Lunge sidewards to the left, swing the rifle forward and upward to the overhead position.

(b) TWO-- Turn the trunk to the left and bend forward over the left hip. At the same time, swing the rifle to a **low** horizontal in front of the left ankle.

(c) THREE-- Recover to position ONE.

(d) FOUR-- Recover to the starting position.

(e) FIVE, SIX, SEVEN, and EIGHT-- Repeat on the right side.

c. Exercise 3: Foreup, Back Bend

(1) Starting Position. Rifle downward, feet together.(See fig. 3-7, C.)

(2) Cadence. Moderate.

(3) Movement. A four-count exercise:at the count of—

(a) ONE-- Swing the arms forward and upward to the overhead position.

(b) TWO-- Bend backward, emphasizing the bend in the upper back. The face is up.Keep the knees straight.

(c) THREE-- Recover to position ONE.

(d) FOUR-- Recover to the starting position.

d. Exercise 4:Up and Forward

(1) Starting PositionRifle downward, feet together.(See fig. 3-7, D.)

(2) Cadence. Fast.

(3) Movement. A four-count exercise:at the count of—

(a) ONE-Swing the arms forward and upward to the overhead position.

(b) TWO- Swing the arms forward to shoulder level.

(c) THREE-Recover to position ONE.

(d) FOUR-Recover to the starting position.

e. Exercise 5: Foreup, Full Squat

(1) Starting Position. Rifle downward, feet in narrow stance. (See **fig.** 3-7, E.)

(2) Cadence. Moderate.

(3) Movement. A four-count exercise: at the count of—

(a) ONE-- Swing the arms forward and upward to the overhead position.

(b) TWO-- Swing the arms down to shoulder level and assume the squatting position.

(c) THREE-- Recover to position ONE.

(d) FOUR-- Recover to the starting position.

f. Exercise 6: Arms Forward, Side Bend

(1) **Starting Position.** Sidestraddle, regular stance, rifle forward.(See fig. 3-7, F.)
(2) **Cadence.** Moderate.
(3) **Movement.** A four-count exercise:at the count of-
(a) **ONE**-- Bend the trunk to the left. Keep the knees straight.
(b) **TWO**-- Recover to the starting position.
(c) **THREE**-- Bend the trunk to the right. Keep the knees straight.
(d) **FOUR**-- Recover to the starting position.
NOTE:Keep the rifle on the same level as the shoulders throughout the exercise.

3303. LOG DRILL

a. Description and Objective. Log exercises are conditioning exercises performed with a log. Each drill contains six exercises and they are numbered in a set pattern. The drill takes 15 minutes to complete. The objective of log drills is to develop strength and muscular endurance and, in this instance, under maximum loads. Log exercises also develop teamwork. Log may be used in lieu of conditioning drills after the Marines have become somewhat conditioned.

b. Area and Equipment. Any level area is satisfactory for conducting this drill. Each Marine completes these exercises from a standing position and no ground contact is required. If the group exceeds a platoon in size, then the instructor will need instructor's stand. Each six-person group or eight-person group will need a log. The logs should be from 6 to 8 inches in diameter. They may vary in length from 14 feet (for 6 people) to 18 feet (for 8 people).The logs should be skinned, smoother, and dried. The 14-foot logs should weigh approximately 300 pounds and the 18-foot logs, approximately 400 pounds. Rings should be painted on the logs to indicate each person's position. When not in use, the logs should be stored on a rack to keep them off the ground.

c. Formation. All the Marines assigned to the same log team should be about the same height at the shoulders. The recommended method of dividing the platoon is to have the Marines form a single file or column with short people to the front and tall people to the rear. Have the Marine assume their positions in the column according to shoulder height, not head height. When they arc in position, they are given the command COUNT OFF BY SIXES (OR EIGHTS), COUNT OFF, to divide them into six- or eight-person log teams. Each team in turn, can then proceed to the log rack, shoulder a log, and carry it to the designated exercise area. The log teams form in columns in front of the instructor. With the Marine holding the log in the chest position, have them face the instructor and ground the log at least 10 yards between columns and 10 yards between log teams within the columns.

d. Starting Positions. The Marines fall in, facing the log, with their toes about 4 inches from it. The basic starting positions and commands are as follows:(See fig. 3-8.)
(1) **RIGHT-HAND STARTING POSITION, MOVE.** It the command MOVE, move the left foot 12 inches to the left, and lower the body into a flat foot squat. Keep the back straight, head up, and arms between the legs. Encircle the far side of the log with the left hand. Place the right hand underneath the log.(See fig. 3-8, A.)

(2) LEFT-HANDED STARTING POSITION, MOVE. These commands are executed in the same manner as in paragraph a. except that the left hand is underneath the log and the right hand encircles its far side. (See fig. 3-8, B.)

Figure 3-8 is available in the illustration section at the end of the book.

(3) RIGHT SHOULDER POSITION, MOVE. At the command MOVE, pull the log upward in one continuous motion to the right shoulder. At the same time, move the left foot to the rear and stand up, facing left. Balance the log on the right shoulder with both hands.(See fig. 3-8, C.)This movement cannot be performed from the left-hand starting position because of the position of the hands.

(4) LEFT SHOULDER POSITION, MOVE. These commands should be given from the left hand starting position. At the command MOVE, pull the log upward in one continuous motion, to the left shoulder. At the same time, move the right foot to the rear and stand facing right. Balance the log on the left shoulder with both hands.(See fig. 3-8, D.)This movement cannot be performed from the right-hand starting position.

(5) WAIST POSITION, MOVE. From the right hand position pull the log waist high. Keep the arms straight and fingers laced underneath the log. The body is inclined slightly to the rear and the chest is lifted and arched. (See fig. 3-8, E.)

(6) CHEST POSITION, MOVE. This command should be given after the waist position has been assumed .On the command MOVE, shift the log to a position high on the chest, bring the left arm under the log and hold the log in the bend of the arms.(See fig. 3-8, F.)Keep the upper arms parallel to the ground.

(7) To move the log from the right shoulder to the left shoulder, the command is: LEFT SHOULDER POSITION, MOVE. On the command MOVE, push the log overhead and lower it to the opposite shoulder.

(8) To return the log to the ground from any of the above positions, the command is: STARTING POSITION, MOVE. At the command MOVE, slowly lower the log to the ground. The hands and fingers must be kept from under the log.

e. Leadership. A principal instructor demonstrates and leads the drill. He must be familiar with the leadership techniques for conditioning exercises and the peculiar techniques for log drill.

3304. EXERCISE PERFORMED WITH LOGS

The exercises of log drill are outlined in the following paragraphs. Figure 3-9 graphically explains log drill.

a. Exercise l:Two-Arm Pushup

(1) Starting Position. Right or left shoulder position. Regular stance. (See fig. 3-9, A.)

(2) Cadence. Moderate.

(3) Movement. A four-count exercise:at the count of—

- **(a) ONE--** Push the log overhead until the elbows lock.
- **(b) TWO--** Lower the log to the opposite shoulder.
- **(c) THREE--** Repeat the action of count ONE.
- **(d) FOUR--** Recover to the starting position.

b. Exercise 2: Forward Bender

(1) Starting Position. Chest position. Regular stance.(See fig. 3-9, B.)

(2) Cadence. Moderate.

(3) Movement. A four-count exercise:at the count of-

(a) **ONE**-- Bend forward at the waist, keeping the back and legs straight.

(b) **TWO**-- Recover to the starting position.

(c) **THREE**-- Repeat the action of count ONE.

(d) **FOUR**-- Recover to the starting position.

c. Exercise 3: Straddle Jump

(1) Starting Position. Right or left shoulder position, feet together, fingers interlaced on top of the log.(See fig. 3-9, **C.)**

(2) Cadence. Moderate.

(3) Movement. A four-count exercise:at the count of—

(a) **ONE**-- Jump to a side-straddle. Pull down on the log with both hands to keep it from bouncing on the shoulder.

(b) **TWO**-- Recover to the starting position.

(c) **THREE**-- Repeat the action of count ONE.

(d) **FOUR**-- Recover to the starting position.

Figure 3-9 is available in the illustration section at the end of the book.

d. Exercise 4: Side Bender

(1) Starting Position. Right shoulder position, feet regular stance.(See fig. 3-9, D.)

(2) Cadence. Moderate.

(3) Movement. A four-count exercise:at the count of—

(a) **ONE**-- Bend sideward to the left as far as possible, bending the left knee.

(b) **TWO**-- Recover to the starting position.

(c) **THREE**-- Repeat the action of count ONE.

(d) **FOUR**-- Recover to the starting position.

(4) After completing the required number of repetitions, change shoulders and execute an equal number of repetitions to the other side.

e. Exercise 5: Knee Bend

(1) Starting Position. Right or left shoulder position. Narrow stance. Fingers interlocked on top of the log.(See fig. 3-9, E.)

(2) Cadence. Slow.

(3) Movement. A four-count exercise:at the count of—

(a) **ONE**-- Flex the knees to a quarter-squat position.

(b) **TWO**-- Flex the knees to a half-squat position.

(c) THREE-- Lower the body to a three-quarter squat position.(Lean slightly forward.)

(d) **FOUR**-- Recover to the starting position.

NOTE:Pull forward and downward on the log throughout the exercise.

f. Exercise 6: Overhead Toss

(1) Starting Position. Right or left shoulder position, regular stance. The knees are bent to a quarter-squat.(See fig. 3-9, F.)

(2) Cadence. Moderate.

(3) Movement. A four-count exercisc:at the count of-

(a) ONE-Straighten the knees and toss the log into the air approximately 12 inches overhead. Catch the log with both hands and lower it toward the opposite shoulder. As the log is caught, lower the body into a quarter-squat.

(b) TWO-- Again toss the log into the air and when caught return it to the original shoulder.

(c) THREE-- Repeat the action of count ONE.

Section IV. Grass Drills

3401. GENERAL

a. Description and Objective. Grass drills are extremely strenuous exercises and are performed at top speed for only short periods of time. No cadence is counted but the Marines continue to execute the multiple repetitions of the command until the next command is given. The grass drill consists of two drills:Drill One and Drill Two. Each drill contains six exercises. The objective of the drills is to decrease reaction time, to develop cardiovascular endurance, and to provide a vigorous workout for all major muscles.

b. Area and Equipment. Any level area suitable for ground contact and of a size to accommodate the group is adequate. No equipment is needed.

c. Formation. All movements are executed in place. The extended, rectangular formation is recommended for a platoon- or company-size unit. The circle formation is suitable for groups of squad or section size. At the beginning of an exercise program, 2 to 3 minutes of grass drills will insure a good workout.

d. Progression. Progression is gained by gradually increasing the length of time devoted to the drills. As the physical condition of the Marines improves, the periods should be gradually lengthened to 5 minutes. As the second drill is more difficult than the first, some progression can be attained by initially executing Drill One; then as the program and the Marines progress, introduce Drill Two. To extend the duration of the drill, it may be necessary to repeat the drill.

e. Starting Position

(1) The drills are started from the GO position. Other basic positions are FRONT, BACK, and STOP. (See fig. 3-10, A.)

(a) GO. Running in place (top speed): on the toes and balls of feet, knees raised high, arms pumping, body bent forward at waist.

(b) FRONT. Prone position: elbows bent (along body), palms flat on ground directly under the shoulders, legs together and straight.

(c) BACK. Supine position (flat on back): arms extended near side on ground with palms down, legs together and straight, feet toward the stand or instructor.

(d) STOP. Football lineman stance: feet spread and staggered, left arm across left thigh, right arm straight, knuckles on ground, head up, back parallel with ground.

(2) To assume the FRONT or BACK position from the STANDING, GO, or STOP position, vigorously get into the prescribed position as quickly as possible. (See fig. 3-10, B.)

(3) To change from the FRONT to the BACK position, quickly do a pushup, move the feet several short steps to the right or left, lift the arm on the side toward which the feet move, and thrust the legs vigorously to the front. (See fig. 3-10, C.)

Figure 3-10 is available in the illustration section at the end of the book.

(4) To move from the BACK to the FRONT position, sit up quickly, placc both hands on thc ground to the right or the left of the legs. Move the feet several short steps to the rear on the side opposite the hands. When the feet are opposite the hands, thrust the legs vigorously to the rear and lower the body to the ground.(See fig. 3-10, D.)

f. Leadership. A warm up activity of lesser intensity should proceed grass drill. During the instructional phase and conduct of these drills, the following points should be applied.
(1) The instructor executes only GO and STOP with the Marines. This allows the instructor to supervise the drill.
(2) The commands for grass drills are given in rapid succession without the usual preparatory command.
(3) To prevent confusion, the instructor should give the commands sharply to distinguish them from comments or encouragement.
(4) As soon as the Marines know the drill, they should respond to the instructor's commands and perform all exercises vigorously and as rapidly as possible. All exercises are executed continuously until the next command is given. Insist on top speed performance; anything less is not effective.
(5) The commands peculiar to each exercise are identical to the name of the exercise.
(6) Marines arc not to be required to assume the position of attention once the drills are started. To halt the drill for instructions or for rest, the command UP is used. At this command, the Marines assume a relaxed standing position. Do not demand formality. At the conclusion of a fast and vigorous 5-minute grassdrill, it is physically impossible for people to stand at attention.
(7) The sequence of commands for the execution of grass drills should occur as follows:Drill One. GO, FRONT, Bouncing Ball; GO, BACK, Bicycle; GO, Full Squatter; GO, BACK, Situps; GO, FRONT, Mountain Climber; GO, FRONT, Roll Left; GO, STOP, UP.

g. Use With Other Programs. Since grass drills can be executed in a short period of time, they may be executed where only a few minutes are available for exercise or in conjunction with another type of activity. Grass drills are an excellent substitute for running when time is a factor.

3402. GRASS DRILL ONE AND TWO

a.Grass Drill One
(1) Bouncing Ball. From the FRONT position, push up, supporting the body on the hands (shoulder-width apart) and feet. (See fig. 3-11.)Keep the back and legs in line and the knees straight. Bounce up and down by a series of short, upward springs from the hands, hips, and feet simultaneously,
(2) Bicycle. From the BACK position, raise the legs and hips. Keep the elbows on the ground and support the hips with the hands. Move the legs vigorously as if pedaling a bicycle.
(3) Full Squatter. From the STOP position, assume a full knee bend, the feet on line, hands on hips. Bounce up and down in place by short, bouncing jumps.

Figure 3-11 is available in the illustration section at the end of the book.

(4) Situps. From the BACK position and with arms stretched overhead, sit up, reach forward, and touch toes. Return to the supine position.

(5) Mountain Climber. From the STOP position, place both hands on the ground directly under the shoulders. Thrust the right leg to the rear, knee straight. The left foot should be close to the left hand, the left knee outside the left arm. Shift the weight to the hands, thrust off with the rear (right) foot and bring that foot up close to the right hand, the right knee outside the right arm. At the same time, thrust the left leg vigorously to the rear, knee straight. Continue at a fast cadence, alternating the legs.

(6) Roll Left. From the BACK or FRONT position, make one complete roll in the direction commanded. On completing the roll, return to the FRONT or BACK position.

b. Grass Drill Two

(1) Legs Over. .From the BACK position and with arms stretched overhead, palms up, raise the legs upward and then swing them backward over the head until the toes touch the ground behind the head. Return legs to the starting position. (See fig. 3-12.)

(2) V-Up and Touch Toes. .From the BACK position, raise the legs with the knees straight, sit up until the trunk and legs form a V, and touch the toes with the hands. Return to the BACK position.

(3) Rocker. .In the FRONT position, clasp the hands behind the back, arch the body, holding the head back. Start rocking, using the front part of the trunk as a rocker.

(4) Bounce and **Clap Hands**. The procedure is the same as for bouncing ball, but while in the air, clap the hands. This requires a more vigorous bounce or spring. (See fig. 3-12.)

(5) Leg Spreader. .From the BACK position, raise the legs so that the heels are 10 to 12 inches from the ground, spread them apart as far as possible, then close them together. Open and close legs as rapidly as possible.

(6) Forward Roll. .For forward roll from the STOP position, place both hands on the ground, tuck the head, and do one complete forward roll, keeping the legs tucked as you roll, and come back to the STOP position.

Figure 3-12 is available in the illustration section at the end of the book.

Section V.Guerrilla Exercise

3501.GENERAL

a. Description and Objective. Guerrilla exercises are individual exercises of various basic skills that are performed rapidly while moving forward in a circle formation. There are two tables of guerrilla exercises. Each table takes 15 minutes to complete. The objective of this exercise is to increase strength and endurance, aid flexibility, and develop coordination.

b. Area and Equipment. Any level area is suitable for conducting guerrilla exercises. No ground contact is required other than the hands. There is no equipment requirement.

c. Formation

(1) The circle formation (app. A) is used for guerrilla exercises. Each platoon forms its own circle and engages in guerrilla exercises under a platoon instructor. If the platoon exceeds 30 Marines, double or concentric circles may be used.

(2) When the circle is formed, the instructor steps into the center of the circle and moves clockwise in a small circle. He commands:QUICK TIME, MARCH, 1-2-3-4. (Rapid cadence of approximately 130 counts per minute. Cadence and step are maintained between exercises.)

(3) To reform the platoon after completing guerrilla exercises, the instructor halts the Marines and places the base man or platoon guide where he wishes and commands:

(a) BASE MAN (or platoon guide),POST.

(b) FALL OUT AND FALL IN ON THE BASE MAN (or platoon guide).

d. Progression. Progression may be attained by moving from table I to table II or by shortening the quick time marching periods between exercises and performing all exercises a second time.

e. Leadership

(1) To execute the exercises, the Marines continue at quick time while the instructor simultaneously explains and demonstrates the exercise to be performed, and then commands the Marines accordingly. In each instance, the preparatory command will be the name of the exercise and, in all instances, the command of execution will ***MARCH***.. *To* terminate each exercise, the command *is* ***QUICK TIME, MARCH.*** *The* Marines immediately pick up the step as the instructor counts cadence.

(2) Unless specified differently, each exercise should be continued for 20 to 40 seconds depending upon the vigor of the exercise. The leader can determine the duration of each exercise by observing its effect upon the Marines.

(3) To form for double guerrillas, the commands for pairing the Marines (who are in circle formation) are-

(a) PLATOON, HALT.

(b) FROM (designate an individual), BY TWO'S, COUNT OFF(.Examplel-2; 1-2; 1-2; etc.)

(c) EVEN NUMBERS MOVE UP BEHIND ODD NUMBERS(At this time, adjust pairs according to height and weight.)

(d) YOU ARE NOW PAIRED UP FOR DOUBLE GUERRILLAS. To change the Marines position, merely command CHANGE.)
(e) FORWARD, MARCH.

f. Place in the Program. Many Marines have not had the opportunity to perform the simple skills involved in guerrilla exercises. The conduct of these exercises is a simple matter since they can be performed easily and quickly in almost any situation. The tables of exercise are applicable to all personnel. The tables can constitute a station within a 1-hour period or be completed within a separate 15-minute period.

3502.GUERRILLA TABLES

a. Table I
(1) Double Time.(See fig. 3-13, A.)Hold arms in the thrust position. Execute a double-time run, maintaining the circle formation and the prescribed distance between your fellow Marines. Duration-1 minute.
(2) All Fours.(See fig. 3-13, B.) Face downward. Support the body with the hands and feet. Walk forward hands first.
(3) Crab Walk.(See fig. 3-13, C.) Get in the sitting position, face upward, and lift the hips. Support the body with the hands and feet. Walk forward feet first.
(4) Squat Walk.(See fig. 3-13, D.) Assume a full knee bend position. Grasp the ankles (left ankle with the left hand, right ankle with the right hand). Walk forward.
(5)Broad Jump.(See fig. 3-13, E.) Jump forward on both feet in a series of broad jumps. Swing the arms vigorously to assist the jumps.
(6)Toe-Touch Walk.(See fig. 3-13. F.) Walk forward, bending at the waist and touching one hand to the toe of the opposite foot while it is on the ground. Raise the trunk to the vertical position between steps. Keep the knees straight.
(7)Bottoms-Up Walk.(See fig. 3-13, G.) Assume the front leaning rest position and move the feet toward the hands in short steps, keeping the knees locked. When the feet are as close to the hands as possible, walk forward on the hands to the front leaning rest position.
(8) Straddle Run.(See fig. 3-13, H.) Run forward, leaping to the right from the left foot and to the left from the right foot.
(9) Fireman's Carry .(See fig. 3-13,1.)
(lO)Single-Shoulder Carry(See fig. 3-13, J.) Two men execute the carries as indicated by the diagram. No. 1 man executes one type; No. 2 man executes the other.

Figure 3-13 is available in the illustration section at the end of the book.

b. Table II
(1) Double Time.(See fig. 3-14, A.)Hold arms at the thrust position. Execute a double-time run, maintaining the circle formation and the prescribed distance between your fellow Marine.Duration-1 minute.
(2) Toe-Grasp Walk.(See fig. 3-14, B.) Bend forward and grasp toes. With knees slightly bent, walk forward.

(3) Hand-Kick Walk..(See fig. 3-14, C.) Walk forward, kicking the moving foot upward on every step. At the same time, lean forward and touch the elevated toe with the hand of the opposite arm.

(4) Pike Juinping.(See fig. 3-14, D.) Jump forward and upward from both feet, keeping the knees straight, and at the same time, swing the legs forward and touch the toes with the hands at the top of each jump.

(5) Squat Jump.(See fig. 3-14, E.) Leap forward from the squatting position, with the hands on the ground and the arms between the legs. Land on the ground with hands and legs extended. Bring up the legs to the squatting position.

(6) Steam Engine..(See fig. 3-14, F.) Lace the fingers behind the neck and walk forward in the following manner: as the left leg moves forward, raise the knee high, bend the trunk forward, and touch the outside of the right elbow to the outside of the knee. Then lower the left leg and step forward on the left foot and raise the right leg. Repeat with the right leg and left elbow.

(7) Knee-Touch Walk.(Scc fig. 3-14, G.) Walk forward, bending the knees and touching the ground on each step. The knees are bent and straightened on each step.

(8)Hobble Hopping.(See fig. 3-14, H.) Hold foot behind back with opposite hand and hop forward. On the command CHARGE, grasp the opposite foot with opposite hand and hop **forward.**

(9) Cross Carry.(See fig. 3-14,1.)(10)Saddle Back Carry. (See fig. 3-14, J.) Two Marines execute the carries as indicated in the diagram. No. 1 Marine executes onetype;No. 2 Marine executes the other.

Figure 3-14 is available in the illustration section at the end of the book.

Section VI. Running and Orienteering

3601. RUNNING

a. General

(l)Description.The general form and technique for all types of running are fairly constant. (See fig. 3-15.)The head is erect, body slightly forward without bending at the waist, and the arms are at a loose thrust position alternating from front to rear in straight planes. A cross-body arm movement wastes energy .The movement of the legs and feet will be discussed in subsequent paragraphs dealing with the different types of running. Of primary importance is the fact that in all types of running, the toes should be pointed straight ahead. Toeing out is a common error in both running and walking and should be an item of individual correction.

(2) Objective. The objective of running is to develop cardiovascular endurance. Despite the fact that Marines have developed their muscle structure and the strength of their muscle tissue, unless they have developed cardiovascular endurance to a satisfactory degree, they are not entirely physically fit or combat ready. Cardiovascular endurance (wind) depends on the efficiency of the lungs and heart. The efficiency of the lungs and heart depend on the amount of oxygen the lungs can absorb with each breath inhaled and the amount of carbon dioxide the lungs can expel. The process of absorbing oxygen and expelling carbon dioxide (cardiopulmonary process) is performed by the blood that circulates through the lungs. The condition of this process will determine the amount of effort a person can exert over a period of time. Running is one of the best activities to develop this vitally important endurance.

b. Running Skills. In the development of running skills, individuals may require instruction to improve their proficiency. Some important skills to consider are-

Figure 3-15 is available in the illustration section at the end of the book.

(1) Action of the Arms. Arm action is important. Check to see that arms are held loosely and that the action is relaxed. The faster the run, the more rapid the arm action.

(2) Breathing. Allow the individual to breathe through the mouth as the body demands a large supply of oxygen. Oxygen can be inhaled in greater quantities through the mouth.

3602. TYPES OF RUNNING

a. Double Time

(1) Description and Objective. Double timing is marching at the rate of 180 steps per minute, each step being 36 inches in length. It takes practice to double time with precision in formation. The Marines should keep in step, placing their feet flat on the ground. This, however, should not be a stamping motion, but should be done with as slight a jolt as possible. Double timing is like a jog, the difference being that in a jog the feet are lifted well off the ground and the running motion is bouncy. In double timing, the feet skim the ground and there is no bounce to the ran. Double timing is a vehicle for teaching proper running form and for the development of the cardiovascular system.

(2) Area and Equipment. This type of running can be completed over a variety of surfaces. Usually a Marine uses a field or road. There is no equipment requirement.

(3)Progression. There is no set standard for alternating quick time and double time in the early conditioning of Marines. A general rule is to begin with enough quick-time marching to ensure a thorough warming up, then double time about 100 paces. Change again to quick time until the individuals have made a reasonable recovery from the running, then double time another 100 paces. The amount of double time can be increased and the quick time decreased from week to week, until the individuals are double timing about 1800 yards. This type of training should be given at least twice a week, but by no means is it adequate as the sole means of conditioning.

(4) Leadership

(a) The instructor should be to one side of the column or group and toward the rear so there is a full view of all Marines. Inexperienced instructors have a tendency to supervise from a position too far forward.

(b) Select an individual who can maintain the proper cadence to act as the guide during double-time running.

(c) There are several ways for the instructor and group to count cadence while double timing. If not contrary to local policy, learn several methods and use them for variety.

b. Wind Sprints

(1) Description and Objective. This type of running involves a series of 30- or 40-yard dashes, usually conducted in successive waves of squads. Each squad is in line and the squad leader on the right flank. Wind sprints assist indeveloping speed and cardiovascular endurance. Any flat and level area may be used which will permit the squad to form a line and run the required distance.

(2) Progression. One or two 30-yard sprints will be adequate at the beginning. As time passes, sprints can be lengthened and up to six or seven sprints may be used.

(3) Leadership

(a) At the command READY (given by squad leader), each runner assumes the sprinter's starting position. At the command GO, the squad sprints approximately 30 yards, takes 10 yards to stop, and lines up immediately with the squad leader who repeats, READY, GO, and again the squad sprints. At the conclusion of the third sprint, the squad waits until all the squads of the platoon have made three sprints. Then they all line up and the squad leaders conduct three more wind sprints in the opposite direction.

(b) Valuable time is gained by having each squad ready to go when the preceding squad has moved off its second sprint mark.

c. Cross-Country Running

(1) Description and Objective. Cross-country running is a distance run conducted on a course laid out along roads, across fields, over hills, through woods, and on any irregular ground. The cross-country ran may be utilized as a conditioner or as a competitive event; the objective is to cover the distance in the shortest possible time. The course should be 2 to 2.5 miles in length and be laid out to avoid heavy vehicular traffic. The course should be marked by directional arrows until the runners know the course. These runs build leg muscles, increase lung capacity, and develop endurance. Any local area of varied terrain is suitable.

(2) Progression. In the mass training of a large group, leaders should be stationed at the head and the rear of the column and should make every effort to keep the individuals together .After determining the abilities of the individuals in cross-country running, it is advisable to divide the unit into three groups. The poorest conditioned group starts first, and the best conditioned group, last. The starting time of the groups should be staggered so that all of them finish about the same time. In preliminary training, the running is similar to ordinary road work in that it begins with rather slow jogging, alternating with walking. The speed and distance of the run is gradually increased. As the condition of the individual improves, occasional sprints may be introduced. At first the distance run is from 1/2 to 1 mile. It is gradually increased to 2 or 2.5 miles. Well-conditioned personnel can run 2 to 2.5 miles within a 15-minute period.

(3) Leadership. Marines should not be required to take part in distance running until they have been through a progressively scheduled training program which requires a considerable amount of running. Cross-country runs should be scheduled occasionally to provide variety in the program. Cross-country running has the advantage of allowing mass participation. Interest can be stimulated by putting the runs on a competitive basis. As a single activity, short cross-country runs can be scheduled once a week, gradually increasing the distance as the physical conditioning improves.

(4)Use With Other Programs. Cross-country running can be combined with other activities such as conditioning exercises.

d. Fartlek Training

(l)Description and Function. Fartlek training is various running exercises conducted along an intense combat-like course. The running is conducted to overwork the lungs, allowing only partial recovery which is followed by another intense period of overload. This sequence is repeated for the duration of the workout. Fartlek training is a useful combat training method and general cardiovascular conditioner. The entire unit must be in uniformly good to excellent condition to qualify for this training.

(2) Progression. The following is an example of one fartlek session:

(a) Warm up by stretching 3-5 minutes.

(b) Warm up running easily 5-10 minutes.

(c) Run at a fast, steady pace for 3/4 to 1 3/4 miles (dependent on terrain).

(d) Walk/jog at a moderate pace for 5 minutes (recovery).

(e) Run easily sprinting 15 to 20 meters occasionally.

(f) Run full speed uphill for 175 to 200 meters.

(g) Warm down by running easily for 1/2 to 1 mile.

(h) Warm down by stretching 3 to 5 minutes.

(3)Leadership. To add variety, try incorporating combat movements, terrain association, a series of exercises or carrying table of equipment weapons (hit and roll). This will take some initiative in setting the course, but it allows for variety. In this manner, the fartlek course can be used as a training session or part of one. lt works extremely well with highly motivated, competitive Marines.

3603.ORIENTEERING

Orienteering is land navigation over a prescribed course as a timed, competitive event. It is an excellent way to integrate land navigation training with physical conditioning and requires Marines to be skilled in both areas. Commanders adjust the training as needed to change the degree of difficulty or to emphasize either navigation or conditioning. Taken to the extreme, an orienteering exercise is conducted in full combat gear over rugged terrain against aggressors. Orienteering requires the following physical skills: endurance, speed, strength, and all-around conditioning.

Section VII. Circuit Training

3701. GENERAL

a. Description and Objective. A strength circuit consists of a series of stations where individuals in small groups exercise vigorously for a short period of time and then move (on signal) to the next station where a different form of exercise is conducted. This rotation continues until all groups move through all stations. Strength circuits contain no set or specific types of exercise stations within the circuit. The objective of circuit training is to develop strength. There are three general types of circuits.

(1) Fixed Circuit. This is a circuit in which apparatus of an immovable type (fixed into the ground) is used.

(2) Movable Circuit. This circuit consists of individual exercise apparatus which is portable and can be moved to and from the training area.

(3) Simplified Circuit. This circuit requires no equipment or apparatus.

b. Formation. The exercises are done at will, but rapid, steady, and continuous work is required of all. Each Marine's nervous and muscular system reacts differently to timed vigorous exercises. His performance should be measured on how many movements per exercise he can complete as an individual. For example, one Marine may be able to complete 5 movements, while another may be able to complete 20, and yet each is receiving the maximum benefit. All three circuits contained in this chapter are designed for platoon-sized groups. Expansion beyond this capacity requires a large amount of equipment, as each Marine in the fixed and movable types of circuits must have an item of equipment available for exercise at each station. A group larger than a platoon could be exercised through use of the simplified type of circuit; however, the group would be unwieldy and control could be a problem.

c. Place in the Program. All circuits illustrated can be completed in a 15-minute period. This feature allows the exercise of a platoonor smaller group on the circuit for a single 15-minute period, or the scheduling of the circuit as a 15-minute period within a longer period. A circuit can thus be utilized within a rotating activity system of scheduling. Choice of a circuit by the unit depends upon area, facilities, and other local factors; however, there is a circuit for every need.

3702. FIXED STRENGTH CIRCUIT

a. Description and Objective. The strength circuit is an arrangement of various types of exercise apparatus which are fixed in position. (See fig. 3-16.)Seven basic exercises are used and each exercise requires an apparatus. All apparatus of one type are positioned together to constitute a station. Each station will accommodate 10 Marines. The circuit is designed to be accomplished in 15 minutes when conducted on a time-rotation basis, as normally executed for unit training. For individual use, a Marine may complete the circuit by executing a specific number of repetitions for each exercise. In this case, the time required for completion of the circuit would vary slightly depending on the number of repetitions the Marine completed. The objective of this circuit is to provide a series of exercises which will improve and maintain the strength of the body's major muscle groups.

Figure 3-16 is available in the illustration section at the end of the book.

b. Starting Level. Marines must be thoroughly warmed up prior to participating in the circuit system. If Marines have not engaged in vigorous exercise immediately prior to starting the circuit, then an instructor should execute the following warm up exercises. These exercises should be conducted in the normal formation for set drills. Seven repetitions of each exercise will normally provide sufficient warm up. The exercises are-

(1) High jumper, Conditioning Drill 1.
(2) Bend and reach, Conditioning Drill 1.
(3) Squat bender, Conditioning Drill 1.

c. Progression. The instructor can adjust the circuit system through several methods to accommodate participants of varying physical ability. One method of adjusting an exercise is to change the method of executing the leg lift or to select a heavier weight for the barbell curls. When it becomes apparent during a unit program that the overall fitness of the group has Improved, then the exercise may be made more strenuous in two additional ways. First, the instructor can have one or more of the supplementary stations added. Second, the time spent exercising at each station can be increased in 5-second increments to a maximum of 60 seconds. Also, as Marines learn the circuit, the 45-second period for movement and instruction between stations can be eliminated, thus allowing only enough time to change stations.

d. Starting Positions. The Marine starts on any station, exercises steadily for a certain period (45 seconds initially), then moves on command to the next type of apparatus where he again exercises steadily for an equal period. The Marine continues until the required exercise is completed at each station. If it is desired to expand the number of stations in the circuit to accommodate more participants at one time, the instructor can provide four additional supplementary exercises, requiring no equipment. This will increase the amount of time required to complete the circuit.

e. Leadership. Close control of all Marines is necessary to ensure that a minimum amount of time is spent in moving them to their initial stations and in moving them between stations. One instructor can control the activity on the strength circuit. A stop watch or wrist watch with a second hand is required. When Marines arrive at the strength circuit, the instructor will form them for exercise and conduct the warm up drill. The group is then reassembled and formed into a number of files equal to the number of stations being used in the circuit. Each file is then directed to a station. As soon as all participants have reached an exercise position at a station, the instructor gives the command READY, GO. After 45 seconds of exercise, the instructor gives the command STOP, CHANGE OVER. The instructor allows Marines 45 seconds for moving to the next station and for preparing for the next exercise before the command EXERCISE is again given. In lieu of verbal commands, a whistle may be used to stop and start the exercises. For large groups, a megaphone or loudspeaker is useful.

f. Individual Conditioning Program. For the Marine working alone on the strength circuit, it would be impractical to time the exercise periods. For individual exercise, the Marine should select a number of repetitions of each exercise to complete, then rotate to the next stations after completing these repetitions. The number of repetitions selected should be at or near the maximum that the Marine is capable of doing without halting for rest.

g. Fixed Circuit Stations

(1) Primary Stations (See fig. 3-16.)

(a) Pullups/Chinups. A horizontal bar placed 8 feet above the ground is necessary for each Marine at this station. The Marine will also need a space on the bar that is 45 inches wide. On the command**EXERCISE**,the Marine grasps the bar with both palms facing either forward or to the rear, arms fully extended, and feet free of the ground and executes the exercise as described in MCO 6100.3_.The exercise is repeated as many times as possible until the command **STOP** is given. Then the Marine moves to the next exercise. If a Marine has done his maximum number of pullups/chinups prior to the command **STOP**, he will remain in the "dead" handing position until the command **STOP** and move to the next station.

(b) Twist Grip. The apparatus is a horizontal bar, free to turn, held between uprights placed 30 inches apart. The bar is 52 inches above the ground. A weight of 20 pounds is attached to the center of the bar by a light rope long enough to permit the weight to rest on the ground. The Marine stands at arm's length from the bar and grasps it with his hands on either side of the rope, palms down, thumbs under the bar. On the command **EXERCISE,** the hands are rotated so that the backs of the hands are rotated away from the body, thus winding the rope on the bar. The elbows are kept straight to ensure that the exercise is performed by the hand and forearm. When the weight is drawn up to the bar, the bar is then rotated in the opposite direction to lower the weight to the ground. This exercise is continued until the command **STOP** is given. The Marine then moves to the next station.

(c) Pulley Weights. The apparatus is a T frame with a system of pulleys that suspends a weight of about 90 pounds. The weight is attached to a light steel cable which has a drawbar attached to the other end. The Marine grasps the drawbar and sits down directly under the bar, legs extended to the front and arms extended overhead. The exercise is executed by pulling the drawbar down behind the head, then extending the arms slowly again until they are fully extended overhead. The exercise is repeated as many times as possible until the command STOP is given. The Marine then will move to the next exercise. Upon completion of the exercise, the weight is lowered slowly to the ground.

(d) Barbell Curls. A barbell is necessary for each Marine at this station. The barbell is constructed of 1 1/4-inch pipe 5 feet long, and two concrete-filled No. 10 cans. Each barbell should weigh about 40 pounds. Variance in the weight of the barbells, up to about55 pounds, will allow appropriate overload to be applied to Marines who are above average in strength or weight. The Marine grasps the bar with the palms forward and assumes a standing position with the barbell held in front of the hips, hands approximately shoulder's width apart. On the command **EXERCISE,** the Marine flexes the elbows and draws the barbell up until it touches the upper chest. The elbows remain at the sides. Breath is inhaled with the upward movement and exhaled as the barbell is lowered to the starting position. The exercise is repeated as many times as possible until the command **STOP** is given. The Marine moves to the next apparatus.

(e) Step-Up. Thc apparatus is a platform or ledge 18 inches high and of such size. At accommodate 10 Marines. The Marine faces the platform and on the command **EXERCISE,** steps up onto the platform, bringing his trailing foot up beside the leading foot. He then steps back down to the original position, stepping down first with the same foot he initially used in stepping up. After 10 repetitions of the exercise, he changes the order of moving the feet to use the opposite leg for stepping up. Repeat this exercise until the command **STOP** is given. The Marine then moves to the next station.

(f) Leg Lift. The apparatus is a horizontal bar constructed as described in (a). To prevent the body from swaying, a horizontal back support is added 40 inches below the horizontal bar. The arms are kept fully extended. On the command **EXERCISE,** the Marine jumps up, grasps the bar with the palms forward and the back support behind him. The exercise is executed by raising the legs to a horizontal position then slowly lowering them to the vertical position. The Marine does not flex his knees. He does not swing the legs to the rear of a vertical position to gain momentum for raising them in the next repetition of the exercise. The movement is repeated until the command **STOP** is given. The Marine then moves to the next exercise. If unable to raise his legs to a horizontal position without flexing his knees, the Marine flexes his knees and draws them up to the chest, then lowers his legs to the vertical position.

(g) Rope Climb. The rope climb is 20 to 30 feet high with five ropes suspended from a horizontal bar which forms the uppermost part of the framework. To prevent the horizontal bar from sagging and to provide safety, only five ropes are attached to it. There are two frameworks per station. The ropes are 6 feet apart. Any method may be used to climb the rope, and the Marines climb as high as possible. Marines who are proficient should climb therope several times during the time allotted. Inexperienced Marines should be cautioned to take care during descent to avoid rope burns on their hands.

(2) Supplementary Stations. The following exercises are designed to expand the basic circuit by being inserted in specific places within the system. For each supplementary station used, there should be adequate room for 10 Marines to exercise.

(a) Bent Knee Situp or Bottoms-Up. These calisthenics are designed to strengthen the abdominal muscles. These exercises will be inserted between the pullup/chinup and twist grip stations. The primary stomach exercise is the situp. In case of inclement weather or other conditions that make ground contact undesirable, the bottoms up exercise is used. In situps on the command **EXERCISE,** the Marine lies on his back (supine position) with knees flexed and both feet flat on the ground and executes the exercise (minus an assistant) as described in MCO 6100.3_.The Marine then returns to the starting position, repeating the exercise until the command **STOP** is given. The Marine then moves to the next station. In bottoms up on the commandEXERCISE,the Marine assumes the front leaning rest and executes the bottoms-up exercise as described in Conditioning Drill 3. He continues this exercise at a moderate cadence until the command **STOP** is given. The Marine then moves to the next station.

(b) Pushup. This exercise is designed to strengthen the arm- and shoulder-girdle muscles. It should be included between the twist grip and pulley weight stations. Upon the command **EXERCISE,** the Marine executes the pushup as described in Conditioning Drill l. He continues this exercise at a moderate cadence until the command **STOP** is given. The Marine then moves to the next station.

(c) Knee Bender. This exercise is designed to build leg muscles and is included between the pulley weight and barbell curl stations. On the command **EXERCISE,** the Marine executes the knee bender as described in Conditioning Drill 3.He continues this exercise at a moderate cadence until the command **STOP** is given. The Marine then moves to the next station.

(d) Trunk Twister. This exercise strengthens the major muscles of the trunk and is included between the step-up and pullup/chinup stations. On the command EXERCISE, the Marine executes the trunk twister as described in Conditioning Drill 1. He continues this exercise at a moderate cadence until the command **STOP** is given. The Marine then moves to the next station.

3703. MOVEABLE STRENGTH CIRCUIT

a. Description and Objective. The exercises in this circuit are progressive and the course is planned to gain and hold the interest of the participating groups. The circuit consists of a series of stations, with each station designed to develop a particular group of muscles. Along with muscular development, correct posture and deep rhythmic breathing should be stressed on this circuit at all times.(See figs. 3-17 and 3-18.)

Figure 3-17 is available in the illustration section at the end of the book.

b. Equipment. The equipment is set up in files. Six files of 8 stations will accommodate a platoon

c. Formation. The platoon marches to the area where the equipment is positioned and forms a file within each lane of stations, covering down on a piece of equipment. Movements are made on the double, the important factor being that no time is wasted in getting to work.

d. Progression. Initially 40 to 45 seconds per station is adequate. As individuals become stronger the time should be increased in 5-second increments until a minute to a minute and a half is reached.

e. Leadership

(1) The leader stands in front of the barbell station and controls the rotation from this position. The leader supervises the entire group, with the assistance of several instructors who move about in the platoon correcting and encouraging the Marines.

(2) The leader starts each group but does not count cadence nor lead them through the exercises. Each individual exercises rapidly but individually.

(3) As each Marine finishes his repetitions with the barbell, he places the barbells on the ground. The leader calls, READY, followed by the command, FALL OUT ONE. All will double-time to the station directly in front of them, while the individuals on the barbell stations will do an about face to the rear station in their lane.

f. Movable Circuit Stations. The Marine can obtain best results on the movable circuit if the exercises on the various stations are given in the following manner:

(1) Station 1—The Barbell. (See fig. 3-17, 1.)The exercises at this station stress proper posture; deep, rhythmic breathing; and development of the muscles of the arms, shoulders, and upper body. An instructor teaches the proper methods of lifting before the exercises begin. The methods are to lift with the legs, to keep the back straight, and to merely grip with the hands. Two recommended exercises are given below. The instructor will use only one exercise per period .He may use either exercise. At successive periods, the other exercise is used.

(a) Exercise 1--Two-Hands Military Press. (See fig. 3-18, 1.)Grasp the barbell with both hands, knuckles up at shoulder width, and lift to the chest. Steadily press to arm's length overhead; lower to the chest resisting weight all the way. Inhale as the weight is pressed up and exhale as the weight is brought down.

(b) Exercise 2--Two-Hands Regular Curl. (See fig. 3-18, 1.) Lift the weight to the waist, with the palms of the hands out, heels together, stomach in, chest lifted and arched, shoulders back, elbows in close to the sides; inhale deeply and curl the weight to the shoulders, using the arms only, at the same time keeping the elbows close to the sides; exhale Emphasize posture and the use of the arms only. This is a very valuable exercise for the development of the biceps and the grip and should be repeated from 8 to 16 times, depending on the ability of the Marine.

(2) Station 2--The Jump Rope. (See fig. 3-18, 2.) This exercise develops strength and agility in the legs and stamina of the whole body. It makes the Marine agile on his feet and increases his footwork efficiency and timing. Each Marine should progress until able to jump rope at least 3 minutes at top speed.

(3) Station 3--The Twist Grip. (See fig. 3-18, 3.) The twist grip is an excellent exercise for the hands and forearms, and adds greatly to the Marine's ability in hand-to-hand combat. The handle is gripped and twisted, winding the rope until the weighted can is level with the height of the hands, which are held horizontal. The weight is lowered in the same manner; the Marine resists the weight all the way, occasionally stopping the twisting motion and alternately removing first one hand, then the other, from the handle. A variation of the above exercise is to wind the handle with the palms up and the arms bent and with the elbows held close in to the sides. Each Marine should maintain a good posture and keep the stomach muscles taut throughout this exercise.

(4) Station 4--The Incline Plane. (See fig. 3-18, 4,)The use of the incline plane is a very strenuous exercise and well designed for the development of the abdomen. Although 6 to 10 repetitions are sufficient for the beginners, more can be added as ability increases. Marines with hernias or recent operations will be excused from participation at this station.

(5) Station 5--The War Club. (See fig. 3-18, 5.) The war club is a simple and effective means of exercising the principal muscle groups of the body, especially those of the trunk, back, and shoulders. To gain the maximum benefit from this exercise, the Marine must keep both feet flat on the ground at all times. Throughout the exercise period, the weight is swung from arm's length as follows:

(a) As in chopping wood, first on one side, then on the other.

(b) As a batter warming up with a number of bats.

(c) In large circles, first with one hand and then with the other.

(6) Station 6--The Bicycle Ride. (See fig. 3-18, 6.) The bicycle ride is well suited to exercising many of the muscle groups of the body, particularly those of the abdomen. Vary the speed of the exercise, but keep everyone "riding" the entire period. A variation exercise may be performed by placing the legs together, raising them slowly to a height about 2 feet from the ground, and then lowering them slowly to the ground.

(7) Station 7--The Step-up. (See fig. 3-18, 7.) The step-up exercises the legs. The step-up is performed by initially stepping up with the left foot, followed by the right, then stepping down with the left foot followed by the right. Continue for 20 seconds, then change to the right foot as the lead foot for 20 seconds.

(8) Station 8--The Isometric Pull. (See fig. 3-18, 8.) Two trainees work at this station with a cable pull and perform the following exercises:

(a) Initially start with one Marine in the supine position and one Marine sitting. The sitting Marine lowers the upper body to the ground and pulls the partner up to the sitting position. The partner then performs this same action and this is continued for 20 seconds at a rapid rate.

(b) During the last 20 seconds, the same action takes place but, in this case, the Marine in the supine position resists the pull of the partner for approximately 5 seconds before allowing to be pulled up into the sitting position.

3704. CIRCUIT-INTERVAL TABLE

a. Description and Objective. The circuit-interval table is designed to develop strength and endurance within a short period of time, with no equipment requirement, through a rapid and vigorous routine of exercise. Fifteen minutes is an adequate period to execute all exercises and to secure a vigorous workout with the circuit-interval principle.

Figure 3-18 is available in the illustration section at the end of the book.

b. Formation and Starting Positions

(1) A leader forms platoons or smaller groups in an oval or circular formation with 3- to 5-yard intervals between Marines. The Marines face to the right and move forward at quick time and then double time. (See fig. 3-19, A.)After running several platoon circle laps, the leader calls out the name of an arm and shoulder exercise from the list below, orders quick time and commands, for example, PUSHUPS. On this command, all Marines immediately hit the ground and individually and rapidly begin doing pushups. No cadence is counted. (See fig. 3-19, B.) After 30 seconds of exercise, the leader commands. **QN YOUR FEET, FORWARD, MARCH** The platoon resumes the quick time cadence and the leader, when ready, gives the necessary commands for double time. The double time is continued for one or more laps and the leader calls out the name of the next exercise and the process is repeated. This continues, with running between each exercise, until every body part has been exercised.

(2) The instructor controls the running and quick time to observe the effects of the exercise upon the Marines. Cadence, step, and precision are not important to the objective and the instructor should not use them. What is important is speed and the instructor should stress this. After the exercise period is started, the Marines do not stop. This circuit method emphasizes stress and recovery, the recovery occurring during the quick time periods.

c. Activities. The leader can use the following exercises. These exercises can be repeated if necessary, during a second round. The leader can use other calisthenic exercises in the circuit-interval table.

(1) Arms and shoulders-push-ups.
(2) Stomach situps.
(3) Back-squat thrusts.
(4) Legs-bicycle (on back).

d. Progression. lhe progress is controlled by the leader, who must pace the running, quick-time movement, and exercise in such a way that Marines will receive a vigorous workout yet be able to participate throughout the 15-minute period. Marines who are in the initial stages of physical condition will not be able to double time or exercise as long as those who are better conditioned. The idea is to set a pace which can be increased during each workout, thus progressing gradually to a higher level of physical fitness.

e. Leadership. The platoon leader, platoon sergeant, or section leader can lead the group. The leader must execute the exercise with the unit to feel the effects and thereby adjust the pace.

f. Place in the Program. This activity may be scheduled whenever a short period of time is available. The only requirement is that enough space, indoors or out, be available to form the circle.

Figure 3-19 is available in the illustration section at the end of the book.

Section VIII. Basic Physical Skills and Obstacle Courses

3801. GENERAL

The purpose of this section is to list the basic military skills and the methods for their development. Many of these skills are best developed by obstacle courses but other drills are also discussed. The section explains types of obstacle courses, details of construction, and methods of negotiating the various obstacles.

3802. BASIC PHYSICAL SKILLS

a. Objective. The objective is to develop proficiency in the various military physical skills which are essential to personal safety and effective combat operations. In travel by foot over rugged terrain and in the execution of combat duties, Marines must be trained to perform basic skills, such as running, jumping, climbing, and carrying. During training, Marines will develop agility and coordination in these skills .Fast and skillful execution of these skills may mean the difference between success and failure on the battlefield.

b. Place in the Program. These skills are practiced throughout the entire program in many different activities. Many of these skills are best practiced on obstacle courses.(See par. 3803.)

c. Description. These basic skills are the minimum skills required by all Marines. The basic skills are as follows:

(1) Running. Running is used to strengthen the legs and develop cardiovascular endurance. Marines should be exposed to running in various situations: on roads, over rough ground, up and down hills, cross-country, and running over low obstacles.

(2) Jumping. In broad jumping, the take-off foot is planted firmly and the spring comes from the extension of this leg as the other leg reaches for the far side of a ditch or similar obstacle. (See fig. 3-20.) The arms are forcibly raised forward and upward to assist in propelling the body up and forward. Landing may be on one or both feet depending upon the length of the jump. In vertical jumping downward from a height, the jumper should aim his feet at the desired landing spot and he should also jump with the knees slightly bent and feet together, with the trunk inclined slightly forward. As the feet touch the ground, the shock is absorbed by bending the knees into a full squatting position. If the height is too great or the ground too hard to absorb the shock, then the jumper should forward roll or side roll thus eliminating some of the momentum.

Figure 3-20 is available in the illustration section at the end of the book.

(3) Dodging In combat situations, it is often times necessary to change directions quickly. To execute this movement while running, a lead foot is firmly planted, left foot if the direction is to the right and right foot if the direction is to the left. The opposite foot is moved toward the new direction. The knees are slightly flexed during the movement and the center of gravity is low and balanced. (See fig. 3-21.) At the time of the change of direction, the head and trunk are turned quickly in the new direction.

Figure 3-21 is available in the illustration section at the end of the book.

(4) Climbing and Surmounting. All Marines should know how to climb and surmount various types of obstacles. (See fig. 3-22.)

Figure 3-22 is available in the illustration section at the end of the book.

(a) Vertical Climbing, as in Climbing a Rope or Pole. The hands grasp the rope or pole overhead with the palms toward the face. Gripping the object, the body is pulled upward with the arms and shoulders, assisted by the feet which grip the object and assist by pushing downward. If shoulder girdle strength and body coordination are not adequate to permit alternating the hands, the arms act together in pulling upward.

(b) Climbing as in Surmounting a Wall. Ingoing over a wall, the body should be kept as close to the top as possible, since in combat operations it is important to offer as small a target as possible to the enemy. If an individual climbs a wall while carrying a rifle, both hands should be freed by slinging the rifle over the back. There are two methods commonly used for surmounting a wall of moderate height, but only one for dropping from it.The methods are as follows:

1 **Run, Jump, and Vault.** Approach the wall at a run, jump forward and upward at the wall and place one foot against it as high up as possible. Use the foot in contact with the wall to help push the body upward while grasping the top of the wall with the hands. Pull the body up with the arms, assisted by pressure of the foot against the wall and swing the legs over, propelling the body weight over the wall.

2 **Hook and Swing.** Approach the wall at a run and jump forward and upward. Hook one elbow over the wall, locking the arm in place by pulling up until the top of the wall is underneath the armpit. Grasp the top of the wall with the other hand. Draw the leg which is closer to the wall up as far toward the top as possible. Then swing the other leg over the top of the wall.The body is then carried over with a rolling motion. A variation of this leg action can be used by Marines who are unable to draw up the leg as described. While hanging with both legs fully extended, start a swinging motion with the legs together. When the legs have enough momentum, swing the outside leg over the top of the wall with a vigorous kick, then follow with the body.

3 **Dropping.** All drops from a height are executed in the same manner regardless of the method used to gain the top. One hand is placed against the far side of the wall while the other hand grasps the top. From this position, the body is rolled over the wall and vaulted away from it with the legs swinging clear. As the body passes over the wall and drops, it should at all times face the wall. This will keep the rifle and other equipment clear. Break the fall by retaining a grasp on the top of the wall as long as possible.

(c) Climbing Ladders and Cargo Nets. Rope ladders, stationary vertical ladders, and cargo nets employ the same general technique. The important element is to grasp the side supports firmly in the hands about shoulder height and place the feet on a rung which would cause the body to be fully extended. In movement upward, one hand is moved upward and a new grasp is secured and, at the same time, the opposite leg moves up a rung. As the knee straightens, the body is elevated. This process is repeated using the opposite arm and leg. Alternation continues in this manner until the climber reaches the objective.

(5) Traversing Horizontal Objects. The traversing of horizontal objects puts heavy stress on the arms and shoulder girdle area as the feet are usually suspended in the air with all of the body weight on the arms and shoulders.(See fig. 3-23.)

(a) Traversing Horizontal Ropes or Pipes. The hands grasp the horizontal support overhead with the palms facing. To propel the body forward, one hand is released and moved forward to secure a new grasp. At the same time, the opposite side of the body is swung forward (some people are able to "walk" in the air, keeping the body to the front and moving the legs in time with the arms as in walking on the ground). The other hand is then released and moved forward; this alternation is continued until the objective is reached.

Figure 3-23 is available in the illustration section at the end of the book.

(b) Traversing Horizontal Ladders. In this situation, the movement is the same as used in traversing a rope or pipe. The hands, however, are placed on the rung with the palms away from the face. Other than this difference the technique is the same.

(6) Crawling. Crawling in combat situations is an often used skill. Crawling may be high or low.(See fig. 3-24.)

Figure 3-24 is available in the illustration section at the end of the book.

(a) High Crawl.

In the high crawl, the Marine moves on hands and knees, moving one hand and the opposite knee and then continuing to move the hands in alternation with the opposite knee following the companion hand.

(b) Low Crawl. The Marine is in the prone position usually with the forearms and palms of the hand on the ground. He propels forward by bending the knee of one leg and pushing with the inside edge of the shoe. At the same time, the opposite arm moves forward and pulls to the rear. The body remains low and movement is continued by bending the opposite knee and pushing, and at the same time sliding the opposite arm forward and pulling. Alternation of hands and legs continues until the objective is reached.

(7) Throwing. Throwing may be executed from the kneeling or standing positions. The object to be thrown is held in the hand, and the throwing arm is bent at the elbow; the hand is then moved to the rear until the hand is behind the ear. The body is turned so that the lead foot and balance arm on the side toward the target are pointing at the target. (See fig. 3-25.) The balance arm is used to sight over and align the throwing hand and the target. When properly aligned, the elbow is moved rapidly forward until it is at a point just in front of the body where the arm is straightened and the wrist snapped. This whip motion propels the object to the target. Underhand throws secure momentum by the thrower bending his knees and swinging the throwing arm to the rear. As the knees are straightened, the arm is forcefully swung forward from the shoulder and the object released.

Figure 3-25 is available in the illustration section at the end of the book.

(8) Vaulting. Vaulting is employed to overcome low barriers or fences. (See fig. 3-26.)The object to be surmounted is approached at an angle. The hand on the side next to the obstacle is placed on the top of the obstacle and, with a straight arm, the body weight is pushed upward. At the same time, the leg on the side next to the obstacle is thrown upward and over the top followed by the other leg (side approach). In landing, the weight comes down on the leading leg first followed by regaining the balance on both legs. The free arm serves as a balance. A direct (front) approach can be used at which time both legs go over the object together.

(9) Man Carrying. There are three basic individual means of carrying personnel in combat situations and one of these methods may be used in carrying objects.

Figure 3-26 is available in the illustration section at the end of the book.

(a) Fireman's Carry."A" stands sideways in front of "B", "A" bends his knees and leans forward, placing one arm through "B's" crotch, grasps the wrist of "B's" arm, which is hanging over the shoulder, and then "A" runs forward.

(b) Saddle-Back Carry. With his back toward "B", "A" stands in front of "B". "B" mounts "A's" hips and clasps his arms in front of "A's" chest. "A" grasps "B's" thighs.

(c) Single-Shoulder Carry. "A" stands facing "B"."A" assumes a semi-squatting position. "B" leans forward until "B" lies across "A's" right shoulder. "A" clasps his arms around "B's" legs and straightens up, lifting "B" from the ground. "A" then runs forward. This method may also be used to carry heavy objects.

(10) Balancing. Balancing the body while walking or running on a narrow object when crossing obstacles is a skill which requires practice and confidence. Balance is required in negotiating a log placed across a stream, in crossing a narrow beam or rail, and in similar situations.(See fig. 3-27.) To perform this skill, place the feet on the object to be crossed, hold the arms to the sides at shoulder level, and fix the eyes on the object approximately 5 yards in front of the feet. Generally, it is not a good practice to look down at the feet. Walk the beam by placing first one foot and then the other in the center of the beam, thereby moving forward, using the arms to aid in maintaining balance.

Figure 3-27 is available in the illustration section at the end of the book.

(11) Falling. Injury can be avoided if Marines are taught to fall properly. They should know how to use the body momentum to their advantage during a fall rather than to try resisting that force. (See fig. 3-28.) If enough force is present, such as occurs during a fall while running or in jumping downward from a height, individuals can extend their hands to catch the weight. At the same time, duck the head and roll forward onto the feet. The key to falling without injury from the standing position is relaxation and rolling on the outside of the leg, hip, and buttocks to take the brunt of the fall.

Figure 3-28 is available in the illustration section at the end of the book.

3803. OBSTACLE COURSES

a. General. Obstacle courses are a valuable part of physical readiness training. The challenge presented by the obstacles assists in developing and testing the basic physical skills. In many combat situations, success will depend upon the Marine's ability to perform one or more of these skills, often while carrying field equipment and when fatigued. In this section, two types of obstacle courses will be discussed-Conditioning Obstacle Course and Confidence Obstacle Course.

b. Course Safety Precautions. Commanders and course instructors should take certain precautions to prevent injury on obstacle courses. A few of the precautions arc:
(1) Inspect the course for faulty construction of obstacles, protruding nails, rotten logs, condition of the landing pits, and other hazards to safety.
(2) Conduct warm up exercises before the unit runs the course.
(3) Explain and demonstrate the correct techniques for negotiating all the obstacles before allowing the Marines to try them.
(4) Give Marines at least two weeks of conditioning exercises before scheduling the obstacle and/or confidence courses.
(5) Ensure that negotiation of the higher and more difficult obstacles are under the supervision of an instructor.
(6) Do not permit individuals who have neither practiced the basic skills nor run the conditioning obstacle course to participate in the confidence obstacle course.
(7) Weather conditions may cause footing or handhold surfaces to be slippery. If such is the case, postpone training on the course.

3804.CONDITIONING OBSTACLE COURSES

a. Description and Objective. The Conditioning Obstacle Course is commonly known as the Obstacle Course. This course consists of fairly low obstacles which are designed to be negotiated quickly. The obstacles serve to test various basic skills, and running the course is a test of the Marine's physical condition. After receiving instruction and an opportunity to practice the skills, Marines run the course against time.

b. Area and Equipment
(1) Complete standardization of obstacle courses should not be attempted since topographical conditions always vary. Commanders should use ingenuity in constructing a course, making good use of streams, hills, trees, rocks, and other natural obstacles. Since the course is eventually run at high speed, it should not be dangerous.

(2) The course should be wide enough for six or eight men to run simultaneously, encouraging competition. The lanes for the first several obstacles should be wider and the obstacles themselves easier than those that follow. This avoids congestion until the contestants scatter out over the course. The last two or three obstacles should not be too difficult and should not involve high climbing. This prevents injuries and falls resulting from fatigue.

(3) The total distance of the course should range from 300 to 450 yards and include from 15 to 25 obstacles. Normally the obstacles should be 20 to 30 yards apart and arranged so that those which exercise the same groups of muscles are separated.

(4) The obstacles should be substantially built. Peeled logs, 6 to 8 inches in diameter, are ideal for many of the obstacles. Sharp points and corners should be eliminated. Landing pits for jumps or vaults should be filled with sand or sawdust to prevent injuries.

(5) The course should be constructed and marked so that it is not possible to sidestep or detour obstacles. However, it is desirable to provide alternate obstacles of varying degrees of difficulty.

(6) The course should be in the shape of a horseshoe or figure eight so that the finish is close to the start and signs should be placed to indicate the course route.

c. Leadership. Before Marines run an obstacle course, they should be instructed in the proper technique of negotiating each obstacle. In each case this technique should be explained and demonstrated in detail, with emphasis on avoiding injury. Every individual should be given an opportunity to practice on each obstacle until he becomes reasonably proficient at negotiating it. Before the course is run against time, it is advisable for individuals to make several runs at a slower pace. During such practice or trial runs, the instructor should observe the performances and make appropriate corrections. Marines should never be permitted to run the course for time until they have practiced on all obstacles. The best method of timing the runners is to have the timer stand at the finish and call out the minutes and seconds as each individual finishes. If several watches are available, each wave may be timed separately. If only one watch is available, the different waves should be started at regular intervals, such as every 30 seconds. If an individual fails to negotiate an obstacle, a previously determined penalty should be exacted.

d.Types of Obstacles

(1) Jumping-Type Obstacles. These obstacles may be ditches which are cleared with one leap, trenches which the individuals can jump into, heights which require jumping downward, or hurdles which an individual can jump over. (See fig. 3-29.)

Figure 3-29 is available in the illustration section at the end of the book.

(2) Dodging-Type Obstacles. Obstacles of this type are usually mazes consisting of posts set in the ground at irregular intervals. (See fig. 3-30.)The intervals between posts should be rather narrow so that the Marines must pick their way carefully through and around them. Lane guides may be constructed to guide the Marines to dodge and change direction. Obstacles may be put into a maze pattern to cause the Marines to change direction.

Figure 3-30 is available in the illustration section at the end of the book.

(3) Vertical Climbing and Surmounting Type Obstacles. These obstacles may be climbing ropes, either plain or knotted and 1-5 inches in diameter; cargo nets or walls 7 or 8 feet high; or vertical poles 6 to 8 inches in diameter and 38 feet high.(Seefig. 3-31.)

Figure 3-31 is available in the illustration section at the end of the book.

(4) Horizontal Traversing-Type Obstacles. Horizontal obstacles may be pipes, beams, ladders, or ropes. (See fig. 3-32.)

(5) Crawling Type Obstacles. Obstacles which require crawling may be constructed of large pipe sections, low rails, and wire. (See fig. 3-33.)

Figure 3-32 is available in the illustration section at the end of the book.

(6) Vaulting-Type Obstacles. Obstacles of 3 to 3.5 feet in height such as low walls or fences may be used as a vaulting obstacle.(See fig. 3-34.)

(7)Balancing-Type Obstacles. Beams, logs, and planks may be used as balancing-type obstacles.(See fig. 3-35.)These items may be used to span water obstacles and dry ditches, or raised off the ground somewhat to simulate natural depressions.

Figure 3-33 is available in the illustration section at the end of the book.

3805.CONFIDENCE OBSTACLE COURSES

a. Description and Objective. The Confidence Obstacle Course is commonly known as the Confidence Course. This course is composed of higher and more difficult obstacles than those used in the conditioning course. The confidence obstacle course is designed to give Marines confidence in their mental and physical capacities and to cultivate their spirit of daring. They are encouraged but not compelled to negotiate this course and the course is not run against time. The negotiation of a confidence course, however, is strenuous enough to be an excellent physical conditioner. Marines should NEVER attempt to take the obstacles at high speed and should not compete for speed. The obstacles vary from fairly easy to extremely difficult. Some are of considerable height to accustom Marines to climbing such heights without fear. Considerable emphasis is placed on obstacles that train and test an individual's balance.

Figure 3-34 is available in the illustration section at the end of the book.

b. Area and Equipment

(1) The confidence course accommodates four platoons, one platoon at each group of six obstacles. The course should be made up of about 24 obstacles, numbered and marked as follows:! to 6, white numbers on red background; 7 to 12, black numbers on a white background;

13 to 18, white numbers on a blue background; and 19 to 24, white numbers on a black background.

(2)A few simple pieces of equipment should be provided for individuals who do not have the strength, courage, or ability to negotiate the obstacles.

Figure 3-35 is available in the illustration section at the end of the book.

c. Formation. The obstacles should be divided into groups of six, and each group is designated by a different color. Each platoon starts at a different color. Individuals are separated into groups of 8 to 12 at each obstacle. At the starting signal from the company commander, they proceed numerically through. Anyone may skip an obstacle who is afraid to try. Individuals proceed from obstacle to obstacle until time is called, then assemble as ordered.

d. Leadership. If the Marines are new to the confidence course, an instructor will demonstrate or will give a brief orientation at each obstacle, including an explanation and demonstration of a method of negotiating it. Marines are encouraged to try the various obstacles, but they are not compelled to do so. No compulsion is to be used. The manner of negotiating any obstacle is left to the discretion of the Marine. However, the instructor assists anyone who experiences difficulty. Instructors must supervise closely at all times to prevent injuries, as some of the obstacles are quite high. Also, some of the obstacles should not be used when slippery or wet. The example of instructors and especially selected demonstrators will serve to inspire the individuals to greater effort.

e. Negotiating the Obstacles. Although personnel need not conform to any one method of negotiating the obstacles, there should be some uniformity in the approach to them. A general method of negotiating the obstacles is indicated below.

(1) Red Group This group contains the first six obstacles. (See fig. 3-36.)

(a) The Belly Buster Individuals may vault, jump, or climb over. Warn them that the log is not stationary.

(b) Reverse Climb. Climb the reverse incline and go down the other side to the ground.

(c) The Weaver. Move from one end of the obstacle to the other by weaving the body under one bar and over the next.

(d) Hip-Hip. Step over each bar, either alternating legs or using same lead leg each time.

(e) Balancing Logs. Step up on log, retaining the balance, walk or run along log.

(f) Island Hopper. Jump from one log to another until the obstacle is negotiated.

(2)White Group. This group is composed of the second six obstacles. (See fig. 3-37.)

(a) Tough Nut. Step over each X in the lane.

(b) Slide for Life. Climb the tower, grasp the rope firmly, and swing the legs upward. Hold the rope with the legs to distribute the weight between them and the arms. Braking the slide with the feet and legs, proceed down the rope. Warn Marines that there is danger of getting rope burns on their hands. When the rope is slippery or wet, this can be a dangerous obstacle.

(c) Low Belly Over. .Mount the low log and jump onto the high log, both arms grasping over the top of the log, the stomach area in contact with it. Swing the legs over the log and lower the body to the ground.

(d) Belly Crawl Move forward under the wire, belly down, to the end of the obstacle.

(e) The Dirty Name Mount the low log and jump to or reach the higher logs in succession, then jump or drop to the ground. Warn the Marines about the height of the final log.

(f) The Tarzan Mount the lower log and walk the length of it and each successive, higher log until reaching the horizontal ladder. Grasp two rungs of the ladder and swing the body into the air. Negotiate the length of the ladder by releasing one hand at a time and swing forward, grasping a more distant rung.

(3) Blue Group. This group is formed by the third group of six obstacles. (See fig. 3-38.)

(a) High Step over. Step over each log, alternating the lead foot or using the same lead foot.

(b) Swinger. Climb onto the swinging log and over to the ground on the opposite side.

(c) Low Wire. Move under the wire on the back, using the hands to raise the wire to clear the body.

(d) Swing, Stop and Jump. Gain momentum with a short run, grasp the rope, and swing the body forward to the top of the wall. Release the rope while standing on the wall and jump to the ground.

(e) Six Vaults. Vault over the logs, using one or both hands.

(f) Easy Balancer. Walk up one inclined log and down the one on the other side to the ground.

(4) Black Group. The last group is formed by the final six obstacles. (See fig. 3- 39.)

(a) Incline Wall. Approach the underside of the wall, jump up and grasp the top and pull the body up and over. Slide or jump down the incline to the ground.

(b) Skyscraper. Jump or climb to the first floor, climb up the corner posts or assist each other to any desired floor. Descend to the ground in any desired manner.

(c) Jump and Land. Climb up the ladder to the platform and jump to the ground.

(d) Confidence Climb. Climb the inclined ladder to the vertical ladder. Go to the top of the vertical ladder, then down the other side to the ground.

(e) Belly Robber. Step on the lower log and assume the prone position on the horizontal logs. Crawl over the logs to the opposite end of the obstacle.

(f) The Tough One Climb the rope or pole on the higher end of the obstacle, then go down the ladder and across the log platform. Climb over or between the logs at the end and go down the rope or pole to the ground. Vault over the final log.

Figure 3-36 is available in the illustration section at the end of the book.

Figure 3-37 is available in the illustration section at the end of the book.

Figure 3-38 is available in the illustration section at the end of the book.

Figure 3-39 is available in the illustration section at the end of the book.

Section IX Individual Exercise Programs

3901.GENERAL

Marines at times will **be** stationed on independent duty and consequently will be responsible for their own physical fitness program. This section will assist them in understanding the need for exercise and will aid in the planning and execution of an individual exercise program. Exercise activities included are the Bench Conditioner, 6-12 Plan, Weight Training, and Isometric Contraction. Keeping physically fit is a problem that faces every Marine. Even though we are frequently engaged in training that requires some physical effort, in many cases, it is not enough to prepare us to meet the intense physical demands of combat. Attaining a satisfactory level of physical readiness is not an insurmountable objective. Available time appears to be the most difficult obstacle to the development of physical readiness. In most cases, regular physical training programs are centralized, requiring the individual to temporarily leave the work area. The problems involved in setting an hour aside two or three times each are numerous. However, most of us can devote a few minutes each day to physical fitness with little, if any, impact on our daily work schedule, especially if it does not require us to leave our work area.

a. Type of Program. There are many good physical fitness programs available to the individual or group. Regardless of the type or duration, to be effective, the program must contain exercises that are strenuous and are challenging to the individual. Space will not permit the inclusion of all available means of individual exercise. The programs selected for this chapter have met the requirement of minimum space and minimum time.

b. Need to Augment Program. These programs are quite strenuous and will develop a satisfactory level of physical readiness. However, if the Marine desires additional development of endurance, it is recommended that he supplement these programs with a 15- to 30-minute period of wind sprints and double timing on an alternating daily basis.

c. Progressive Training. If Marines are performing duties which require little or no physical activity, they must plan a physical conditioning program that assures a moderate beginning, moderate but steady progression, and sufficient warmup before starting the vigorous exercise. To avoid organic of bodily harm, a Marine activity without adequate warmup. He should conduct conditioning programs on a daily basis over an extended period of time, never on an unduly accelerated or crash basis.

3902.THE BENCH CONDITIONER

a. Description and Objective. The bench conditioning program uses a modified bench to employ both isotonic (moving) and isometric (stationary) exercises are as the nucleus of the program. The exercises are designed to develop strength and endurance in all the major muscle groups of the body. The principles of progression, overload, and balance are employed when the exercises are performed properly.

b. Area and Equipment. Thc conditioning apparatus can be constructed in any unit motor pool with welding equipment found in most salvage yards.(See figs. 3-40, 3-41.)Additionally, there are available any number of commercially produced apparatus available. The important thing is that in utilizing this equipment the following exercise routine be adhered to.

c. Starting Level and Progression.. The program consists of two tables, each with 10 exercises. The Marine controls progression by required repetitions or, in some cases, by application of maximum effort. Each table can be completed within 15 minutes.

d. Starting Position. To start the program, the Marine begins with the Table 1 and executes each exercise for the required number of repetitions as indicated. The Marine controls the starting level and progression. When the Marine executes the maximum repetitions for Table I within a 15-minute exercise period, he progresses to Table II. To maintain this level of development, the Marine should also execute the maximum repetitions for Table II within a 15-minute exercise period. The Marine should keep substitution of exercises to a minimum. However, it he completes a full 15 minutes of strenuous exercise and exercises all muscles, then there should be no appreciable difference in the overall development.

e. Bench Conditioning, Table I

(1) Exercise 1: Side-Straddle Hop. This is a two-count warmup exercise done at moderate cadence. The starting position is the position of attention. On count ONE jump slightly into the air, swinging the arms out to the sides and up to a vertical position, hands touching.(See fig. 3-40, A.)At the same time, spread the feet wider than shoulder-width apart. On count TWO, using a slight flexing of the knees and ankles, jump slightly into the air and return to the starting position by swinging the arms back down to the sides. Twenty repetitions of this exercise is the standard dosage throughout the program.

(2) Exercise 2: Hand Walk. Remove the lower horizontal bar. Adjust the upper horizontal bar so that it is high enough to permit a "dead hanging" position, with the feet off the ground.(Sec fig. 3-40, B.j From the "dead hanging" position, release one hand and drop the arm to the side of the body. Then raise that arm and regrasp the horizontal bar. Release the bar with the other hand and drop that arm to the side. Repeat this as many times as possible.

(3) Exercise 3: Situps. Lie down with the fingers interlocked and placed behind the head. Hook the toes under the foot braces. Raise the trunk and upper body to an upright sitting position, twisting it to the left and then forward and downward until the right elbow touches the left knee.(See fig. 3-40, C.)Lower the body to the starting position .Sit up again but twist the body to the opposite direction as before, touching the left elbow to the right knee. Again lower the body to the starting position. The starting dosage is 20 situps. The Marine should continue the progression until he has attained 40 situps.

Figure 3-40 is available in the illustration section at the end of the book.

(4) Exercise 4: Double Step-upStarting at one end of the bench, step up onto the bench, and walk across it. Step down from the other end; turn around and repeat the process to return to the starting point.(See fig. 3-40, D.)Each return to the starting point constitutes a repetition. The starting dosage is 20 repetitions. Maximum dosage is 35 repetitions. This exercise should be done at a rapid cadence.

(5) Exercise 5: Isometric Bar Lift. Adjust the lower bar so that it is slightly higher than the beltline. Placing the feet on the footplates at the base of the frame, grasp the lower bar so that the hands are spread shoulder-width apart. Assume a crouched position and lift with maximum effort using the arms, back and legs.(See fig. 3-40, E.)Starting dosage is 4 repetitions of a stress time of 5 seconds followed by a 5-second rest prior to the next repetition. The Marine obtains progression by lengthening stress periods to 6 and later 7 seconds. Do not increase the number of repetitions.

(6) Exercise 6: Knee Lift. Adjust the upper bar to the same height used in Exercise 2.Adjust the lower bar so that is stops rearward movement of the hips when the "dead hanging" position is assumed.(See fig. 3-40, F.) Keeping the arms extended, flex the legs and raise the knees as high as possible. Hold this position for 5 seconds, then return to the starting position. After 2 seconds in the starting position, raise the knees again. Each return to the starting position constitutes one repetition. The dosage is five repetitions. Progression is the same as in Exercise 5.

(7) Exercise 7: Isometric Pull. Adjust the lower horizontal bar to a position where it is slightly higher than the beltline. Grasp the handles and pull outward.(See fig. 3-40, G.)Apply maximum effort and hold for approximately 5 seconds. Relax for 5 seconds between repetitions; perform four repetitions. Moving the body closer to or farther away from the bar will change the stress from the upper arms to the forearms. Progression is the same as in Exercise 5.

(8) Exercise 8: Isometric Compression. Maintain the position as in exercise 7.(See fig. 3-40, H.)Grasping the handles in the same manner, press in **with** maximum effort and hold for approximately 5 seconds. Relax for 5 seconds between repetitions. Progression is the same as in Exercise 5.

(9) Exercise 9: Isometric Press. Remove the lower horizontal bar. Adjust the upper horizontal bar until it is about 6 inches lower than the extended arms can reach. Stepping on the footplates at the bottom of the frame, grasp the bar with both hands and push up.(See fig. 3-40,1.)Keep both the legs and arms slightly flexed and the back straight. Apply maximum effort for 5 seconds then relax for 5 seconds. Complete four repetitions. Progression is the same as in Exercise 5.

(10) Exercise 10: Pushups. Grasping the foot braces with both hands, assume the front leaning rest position.(See fig. 3-40, J.)Keeping the back and legs straight, lower the body until the chest is lower than the **hands,** then return to the starting position. The Marine should complete the maximum possible number of repetitions.

f. Bench Conditioning, Table II. There is no limit on the maximum number of repetitions attainable in Exercises 3, 4, and 6 of Table II. The only limit imposed is that the entire program of 10 exercises outlinted in either table should not exceed 15 minutes.

(1) Exercise 1: Side-Straddle Hop. This is a two-count warm up exercise done at a moderate cadence. The starting position is the position of attention. On count ONE, jump slightly into the air, swinging the arms out to the sides and up to a vertical position, hands touching.(See fig. 3-41, A.)At the same time, spread the feet wider than shoulder-width apart. On count TWO, using a slight flexing of the knees and ankles, jump slightly into the air and return to the starting position by swinging the arms back down to the sides. Twenty repetitions of this exercise is the standard dosage throughout the program.

(2) Exercise 2: Pull-up. Adjust the horizontal bar so that is is high enough to permit a "dead hanging" position with the feet off the ground. Grasp the bar with both hands, palms facing forward. By flexing the arms, raise the body to a position where the chin is higher than the bar.(See fig. 3-41, B.)Then lower the body to the "dead hanging" position. Repeat as many times as possible.

(3) Exercise 3: Bench situps. Sit on the bench and hook the feet under the foot braces. With the fingers interlocked behind the head, lean back until the head touches the floor.(See fig. 3-41, C.)Return to the starting position. The starting dosage is 15 situps.

(4) Exercise 4: Step-Up. Face the bench and step up on it with one foot, bringing the trailing foot up next to the leading foot. Step back down again, leading with the same foot used first in stepping up.(See fig. 3-41, D.)Perlorm half of the total repetitions, then change the sequence of moving the feet to use the other leg in stepping up, and repeat the same amount of exercise. The starting level is a total of 40 step-up. **This exercise should be done at a rapid cadence.**

(5) Exercise 5: Isometric Bar Lift\djust the lower bar so that it is slightly higher than the beltline. Placing the feet on the footplates at the base of the frame, grasp the lower bar so that the hands are spread shoulder-width apart. Assume a crouched position and lift with maximum effort using the arms, back, and legs.(Sec fig. 3-41, E.)Starting dosage is four repetitions of a stress time of 8 seconds followed by a 5-second rest prior to the next repetition. Progression is obtained by lengthening stress periods to 10 seconds. Do not increase the number of repetitions.

Figure 3-41 is available in the illustration section at the end of the book.

(6)Exercise 6: Leg Lift. Adjust the bars and assume the starting position as shown in figure 3-41, F. Keeping arms and legs extended, raise the legs to a horizontal position and hold in that position for 2 seconds. Then lower the legs slowly to the starting position. Five repetitions is the starting level.

(7) Exercise 7: Isometric Pull. Adjust the lower horizontal bar so that it is slightly higher than the beltline. Grasp the handles and pull outward.(See fig. 3-41, G.)Apply maximum effort and hold for approximately 8 seconds. Relax for 5 seconds between repetitions; perform four repetitions. Moving the body closer to or farther away from the bar will change the stress from the upper arms to the forearms. The Marine obtains progression by lengthening the stress period to 10 seconds.

(8) Exercise 8: Isometric Compression. (See fig. 3-41, H.)Maintain the position as in exercise 7.Grasping the handles in the same manner, press in with maximum effort and hold for approximately 8 seconds. Relax for 5 seconds between repetitions; perform four repetitions. The Marine obtains progression by lengthening the stress period to 10 seconds.

(9) Exercise 9: Isometric Pres. Remove the lower horizontal bar. Adjust the upper horizontal bar until it is about 6 inches lower than the extended arms can reach. Stepping on the footplates at the bottom of the frame, grasp the bar with both hands and push up.(See fig. 3-41,1.)Keep both the legs and arms slightly flexed and the back straight. Apply maximum effort for 8 seconds, then relax for 5 seconds. Complete four repetitions. The Marine obtains progression by lengthening the stress period to 10 seconds.

(10) Exercise 10: Inclined Pushup. Assume the front leaning rest position with the feet on the bench.(See fig. 3-41, J.)Keeping the back and legs straight, lower the body until the nose touches the ground. By extending the arms, raise the body to the starting position. Repeat as many times as possible.

3903. THE 6-12 PLAN

a. Description and objective. The 6-12 plan of physical fitness has been developed to assist in regulating quantity and progression and to provide a convenient set of exercises. This is a basic program and will take 18 weeks to complete if you follow the moderate progression as to the time prescribed for each level of achievement. The time can be shortened as explained below. This plan consists of six basic exercises a day which can be completed in 12 minutes. There are six tables of six exercises each, thus allowing you to progress from table to table. The plan is progressive, fits any age group, contains balance and variety, and applies the principle of overload in a safe and gradual manner. Begin at Table I, Progression Guide, with the number of repetitions as indicated by age.

b. Formation. If just starting an exercise program, do not rush through the first table. Remember, individuals should remain at each level for about a week before moving upward. The time allotment stated for each exercise at the bottom of the tables is a guide; some people may take more and some less time on the individual exercises. At the end of a one-week period, when the individual can comfortable perform the six exercises in 12 minutes, move on to the next level. To a certain degree, the individual must be the judge of his ability to progress form level to level and table to table. If starting with a certain degree of fitness, some of the beginning tables may present little challenge.

c. Starting Level **and Progression.** There are three levels of achievement for each age group, indicated as A, B, and C. Start at the C level for the appropriate age group. At the end of a one-week period, or when the individual can do all exercises at that level within 12 minutes, progress to the B level. At the end of the second week, or when the individual can accomplish that level within 12 minutes, progress to the A level. At the conclusion of the third week or when the individual can achieve the A level within the time limitation, move on to table II.(See Tables I through VI, Progression Guide.)

d. Maintenance Level. Attempt to work through all six tables. If this proves to be too difficult, then maintain exercise at the—

(1) A-level on Table IV, Progression Guide it in the 45 to 49, 50 to 59, or over 60 age group.

(2) A-level on Table V, Progression Guide if in the 17 to 29, 30 to 39, or 40 to 44 age group.

e. Precautions. To achieve the maximum benefit, perform each exercise exactly as specified. Read the descriptions and study the illustrations. Do not slight the movements. Use a sensible approach and follow these four points as they apply before starting or during your exercise program.

(1) If you have the slightest doubt about your ability to participate in this exercise program, consult a physician.

(2) Stop immediately if you notice unusual breathlessness or chest pain while taking part in these exercises. If these conditions persist, consult a physician.

The following Tables of Progression Guides and Exercises are the progression of thc 6-12 Plan program.

Table I. Progression Guide is available in the illustration section at the end of the book.

Table I. Exercises.is available in the illustration section at the end of the book.

Table II. Progression Guide.is available in the illustration section at the end of the book.

Table II. Exercises.is available in the illustration section at the end of the book.

Table III. Progression Guide.is available in the illustration section at the end of the book.

Table III. Exercises.is available in the illustration section at the end of the book.

Table IV. Progression Guide.is available in the illustration section at the end of the book.

Table IV. Exercises.is available in the illustration section at the end of the book.

Table V. Progression Guide.is available in the illustration section at the end of the book.

Table V. Exercises.is available in the illustration section at the end of the book.

Table VI. Progression Guide.is available in the illustration section at the end of the book.

Table VI. Exercises.is available in the illustration section at the end of the book.

(3) Unless you have exercised regularly and know yourself to be in good physical condition, start at table I with the C level appropriate to your age.

(4) If you are out of shape, admit that fact to yourself. Hide your pride; after all, you are in the privacy of your own quarters. Set your goal for the longer, steadier pull toward fitness. Resist the urge to pass over the lower numbered tables to find a table that will test your fitness. You are not trying to test, but rather to develop.

3904.WEIGHT TRAINING

Weight or barbell training should not be confused with the more common types of weight lifting used as a competitive sport. Weight lifting is designed to develop specific muscle groups so that the individual is capable of lifting a large amount during a single lift. In contrast, weight training is the systematic development of all the major muscle groups by the use of calisthenics reinforced with weight to provide resistance.

3905.WEIGHT LIFTING

a. Description and Objective. The weight lifting program is progressive and applies the principle of overload in a safe, gradual manner. The exercises of the table can be completed in 15 minutes. The objective of the exercises is to develop strength and muscular endurance, and muscle tone of the five major muscle groups:legs, arms, back, trunk, and shoulder girdle.(See fig. 3-42.)

b. Warmup. A warm up exercise is important to prepare the body for the more vigorous exercises that are to follow. Ten repetitions of the high jumper exercise are excellent for a warm up period. Also, an exercise which will require fast body movement is needed.To provide such exercise, 3 to 5 minutes of rope skipping is recommended to increase the individual's development potential.

c. Formation. Marines should take care in the completion of these weight lifting exercises. They should ensure that their back is straight during the lifting phase of all exercises. When exercises require assuming the standing position with the weight, they should always grasp the weight while in a squatting position and then rise to a standing position.

d. Progression. Each exercise has a starting number of repetitions and specified pounds of weight. After each 4th or 5th day of exercise, the Marine should increase the repetitions by one until he has reached the maximum of 10 repetitions. At this time, the Marine should increase the weight by 5 pounds. Then he should repeat the process again with the initial number or repetitions.

e. Weight Training

(1) Exercise l:Squat(See fig. 3-42, A.)Starting level—six repetitions, 50 pounds (commonly called the flatfoot deep knee bend). Place the bar upon the shoulders. Stand with feet about 18 inches apart. Keeping the feet flat, lower the body into the low squat position. Stand erect and repeat. Exhale while lowering into the squat position and inhale while standing. This constitutes one repetition.

(2) Exercise 2:Waist Bender(.See fig. 3-42, B.)Starting level-6 repetitions, 40 pounds. Assume the standing position with the bar across the shoulders and the feet a shoulder-width apart .Bend forward at the waist until the upper body is parallel to the ground; return to the starting position. Each return to the upright position constitutes one repetition.

(3) Exercise 3:Curl (See fig. 3-42, C.)Starting level—6 rcpctitions, 40 pounds. Grasp the barbell with the palms facing to the rear and assume the standing position with the feet a should-width apart. With the barbell held in front of the hips, flex the elbows and lift the weight until the bar touches the upper chest. Lower the barbell back to the hip level position. Inhale deeply with the upward movement and exhale on the downward movement. Each time that the bar touches the chest will constitute one repetition.

(4) Exercise 4:Side Bender(.See fig. 3-42, D.)Starting level—6 repetitions per side, 40 pounds. Assume the standing position, with the bar across the shoulders, with feet a shoulder-width apart. Bend to the left as far as possible and return to the starting position. Repeat six times and then execute the same procedure to the right for six repetitions.

(5) Exercise 5:Standing Press(See fig. 3-42, E.)Starting level—6 repetitions, 45 pounds. Grasp the bar with the palms facing forward and assume the starting position. Curl the weight to the upper chest position. Inhale deeply and press the bar upward to an overhead position. Exhale while lowering the bar to the chest position. Each time that the bar is pressed upward constitutes one repetition.

(6) Exercise 6:Upward Row. .see fig. 3-42, F.)Starting level—6 repetitions, 40 pounds. Grasp the bar, hands close together, palms to the rear, and assume the standing position. Starting with the bar held in front of the hips, flexing the elbows and the shoulder girdle muscles, lift the bar straight up to an overhead position. Inhale deeply while lifting the bar. Exhale while lowering the bar to the hip position. Each time that the bar returns to the hips constitutes one repetition.

(7) Exercise 7:Shoulder Curl. See fig. 3-42, G.)Starting level—6 repetitions, 25 pounds. Grasp the bar, palms down, and assume the standing position. Keeping the elbows locked, curl the bar, pivoting the arms at the shoulders until the bar is in an overhead position and as far to the rear as possible. Return the bar in the same manner to the hip position. Each time that the bar returns to the hip position constitutes one repetition.

Figure 3-42 is available in the illustration section at the end of the book.

3906.ISOMETRIC EXERCISING

a. Description and Objective. Isometric exercising is the application of maximum effort during an exercise period. The isometric principle is to apply force gradually over a 5- to 10- second period until the maximum effort is applied. Relaxation follows for approximately 5 seconds and then force is applied again. The Marine continues this process at the prescribed level for each exercise for a period of 15 minutes or less. The objective of isometric exercises is to create muscle growth. Isometric exercises are the fastest means of creating muscle growth. They are founded on the fact that a muscle will grow only so fast regardless of the type or duration of the activity. The principle of overload—that the muscles develop commensurate with demand—reinforces that fact. However, isometrics will not develop cardiovascular or muscular endurance. Consequently, Marines who choose to use isometrics in their exercise routine must also include running or some other type of aerobic activity.

b**. Area and Equipmen**t isometric exercises may be designed to be performed with or without equipment.

c. Use With Other Programs. The isometric exercises presented in this section will provide a variety from which to choose. In addition to the exercises contained in this section, the application of isometric force is used in some of the exercises in section II.

d.Door Frame Exercises. The following exercises are designed for use with a standard door frame found in all offices or barracks.(See fig. 3-43.)

(1) Exercise l:Arm Press. Stand in the doorway with the legs straight, knees locked. Using the arm muscles, press hard upward against the top of the door frame. Repeat for three repetitions applying gradual effort to maximum contraction.

(2) Exercise 2:Leg Press. Stand in the doorway with the hands on the top of the door frame, elbows locked .With knees bent, press hard with the leg muscles .Repeat for three repetitions beginning with a gradual effort and increasing to maximum contraction. A low platform may be necessary to reach the top of the door frame and still maintain a bent knee position.

(3) Exercise 3:Side Press. Extend both arms to the side of the doorway. Palms are shoulder high, facing outward. With both arms, press hard against the sides of the door frame. Repeat for three repetitions. Begin gradually and increase to maximum contraction.

(4) Exercise 4:Lateral Raise. Extend both arms to the sides of the doorway, arms down, palms facing inward. With the back of the hands, press hard against the sides of the door frame .Repeat for three repetitions. Begin with a gradual effort and increase to maximum contraction.

Figure 3-43 is available in the illustration section at the end of the book.

(5) Exercise 5:Neck Press?lace the forehead against the door frame, hands clasped behind the back Using the neck muscles, press hard against the door frame. Repeat for three repetitions, then reverse position so that the back of the head is resting on the door frame. Again do three repetitions .Begin gradually with both exercises and increase to maximum contraction.

(6) Exercise 6:Door Pull Stand facing the edge of the open door and grasp the doorknobs. Pull outward with both arms (if doorknobs are not available grasp the edge of the door).Applying outward pressure, move the body toward and away from the door .Repeat the three repetitions. Begin with gradual effort and increase to maximum contraction.

Chapter 4
COMBAT WATER SURVIVAL
Section I. Marine Corps Water Survival Program

4101.MILITARY SWIMMING

Military swimming emphasizes strokes that result in staying power rather than speed. Marines need to swim easily, quietly, and with adequate vision. Strokes should be used that allow the Marine to carry basic combat equipment and to tow or push a wounded buddy while keeping the face out of the water to allow breathing. The best strokes to use are the side stroke and the breast stroke. Marines must be proficient in floating, drown-proofing, and in using their basic combat equipment to make flotation devices. They must be drilled on using the Kapok and Mae West buoyancy compensators and on emergency egress from helicopters, landing craft, and amphibious assault vehicles.

4102.PROGRAM DEVELOPMENT

Commanders of the Landing Force Training Commands (Atlantic/Pacific) are tasked with developing and maintaining the individual water survival and swimming training program. They are responsible for training all Marine Corps water safety and survival instructors and integrating this training in to their combat readiness training program. All Marines are deployable and as soldiers of the sea must constantly practice the skills necessary to achieve mission accomplishment. Commanders must be imaginative in program development, integrating basic skills development which enhances the total mission accomplishment. An example is teaching waterproofing of the Marine's backpack so that it can be used as a flotation device in streams or rivers. Once learned, this skill would be practiced first in a pool and then in a stream or a river. Routine inspections during deployments would ensure that backpacks could be used as flotation devices.

4103.QUAL STDS AND TRAINING GUIDELINES

The inherent nature of Marine Corps operations and training requires that Marines achieve an ability to survive in water. Water survival and swimming training is designed to reduce fear of water, instill self-confidence, and develop a Marine's ability to survive in water. As a minimum, every Marine should be qualified as swimmer, third-class (S3).This qualification should be met during recruit training or Officer Candidate School. Once qualified, a Marine need not be requalified unless requalification at a specific level (S3 or higher) is required by unit mission, military occupational specialty or duty assignment, or for other reasons determined by the commander. The following classifications and standards apply to Marine Corps water survival and swimming training. Abbreviations will be used for service, medical, and training record entries citing MCO 1510.29A, Individual Water Survival and Swimming Training, as authority.

a. Qualification standards and test procedures are described in paragraph 4104.

b. Swimmer classification and abbreviations are:

(1) Unqualified - UQ.

(2) Swimmer, third-class (S3).Minimum aquatic skill level for all Marines.

(3) Swimmer, second-class (S2)-

(a) Minimum standard for all naval aviators, naval flight officers, and aircrewmen with written waiver from the commanding officer.

(b) Requirements for military occupational specialty 1803 (assault amphibian vehicle officer), 1833 (assault amphibian vehicle crewman), 1302 (engineer officer), 1371 (combat engineer), 1381 (shore party specialist), and 1379 (engineer operations chief).

(4) Swimmer, first-class (S1)-

(a) Naval aviator, naval flight officer, and aircrewman qualification requirement.

(b) Requirements for military occupational specialty 0321 (reconnaissance man), 8652 (reconnaissance man, parachute jump qualified), 8653 (reconnaissance man, self-contained underwater breathing apparatus qualified), 8654 (reconnaissance man, parachute/self-contained underwater breathing apparatus qualified), 9952 (self-contained underwater breathing apparatus Marine (officer/enlisted)), and 9953 (parachutist/self-contained underwater breathing apparatus Marine (officer/enlisted)).

(5) Water Survival Qualified (WSQ).Ultimate water survival/swimming goal for Marines.

c.Instructor and assistant instructor classifications are-

(1) **American Red Cross Advanced Lifesaver.** Individuals who are currently certified as American Red Cross advanced lifesavers may assist qualified instructors during Marine Corps water survival/swimming training and testing. Certification is valid for 3 years.

(2) American Red Cross Water Safety Instruction. Individuals who are currently certified as American Red Cross water safety instructors may conduct swimming training and qualify individuals as 3d, 2d, and 1st class swimmers. Certification is valid for 2 years.

(3) United States Marine Corps Water Safety and Survival Instruction. Those personnel certified as water safety and survival instructors by one of the landing force training commands may conduct Marine Corps water survival and swimming training and American Red Cross swimming and water safety training. These instructors may also participate as assistant instructors during the conduct of landing force training command-sponsored water safety and survival instructor courses. Certification is valid for2 years.

(4) American Red Cross Water Safety Instructor Traine. Individuals qualified who have been certified as instructor trainers by the American Red Cross may train American Red Cross water safety and basic swimming instructors. Certification is valid for 2 years.

(5) United States Marine Corps Water Safety and Survival Instructor Traine. Individuals qualified who are certified as instructor trainers through special training courses conducted jointly by the American Red Cross and the Marine Corps, under the direction of the Commandant of the Marine Corps, may train water safety and survival instructors and American Red Cross water safety and basic swimming instructors. Certification is valid for 2 years.

d. Training Guidelines Instructors engaged in training Marines as swimmers will use the current edition of the American Red Cross swimming and water safety courses instructors' manuals as the authoritative textbooks for swimming procedures, strokes, breaks, and holds.

4104.WATER SURV/SWIM QUAL STD AND TEST PROCS

a. The following qualification standards and test procedures are applicable to Marine Corps water survival and swimming training. The utility uniform without cover, boots/oxfords or socks will be worn for all swimming tests through swimmer, first class. The utility uniform with boots/oxfords will be worn as prescribed for the water survival qualification.

(1) Swimmer, Third-Class (S3) To qualify as a swimmer, third-class, a Marine must enter the water feet first from a minimum height of 5 feet and remain afloat for 5 minutes. During this time the Marine must swim 50 yards using any stroke or combination of strokes. This test should be taken by all Marines as early as possible in their initial training. Those unable to pass this test will be classified as UQ and should be given instruction in fundamental swimming skills. Those who barely meet requirements are swimmers who need help and should receive additional instruction in fundamental swimming skills, whenever possible.

(2) Swimmer, Second-Class (S2.)To qualify, a Marine must enter the water feet first from a minimum height of 10 feet and remain afloat for 10 minutes. The Marine must swim 100 yards and use the three basic survival swimming strokes (side stroke, backstroke, and breast stroke) for a minimum distance of 25 yards each.

(3) Swimmer, First-Class (S1)A prerequisite to qualification as swimmer, first-class, is successful completion of the test for swimmer, second-class. To qualify as a swimmer, first-class, a Marine must be able to do each of the following:

(a) Approach a person of approximately the same size while in the water, demonstrate a release, get that person in a carry position, and tow the "distressed swimmer" 25 yards.

(b) Enter water feet first and immediately swim underwater for 25 yards. Swimmer is to break the surface twice for breathing during this distance at intervals of approximately 25 feet.

(c) Remove trousers in water, inflate for support, and remain motionless for a minimum of 1 minute.

(d) Swim 220 yards using any survival swimming stroke or any combination of survival strokes.

(4) Water Survival Qualification (WSQ)A prerequisite to water survival qualification is successful completion of the test for swimmer, first-class. To be water survival-qualified, a Mannemust:

(a) Enter water by jumping from a height of 10 feet.

(b) Stay afloat in full utility uniform (less cover) for 1 hour. Boots/oxfords will be removed after 5 minutes in the water but will be retained by the swimmer.

(c) Traverse 75 yards in deep water in full utility uniform (less cover) with boots/oxfords.

b. All water survival and swimming training will include instruction in the following:

(1) Employment of safety lines or other comparable expedients to assist in swift stream/river crossings.

(2) Adverse physiological effects caused by cold water (hypothermia) and the precautionary measures to be taken prior to exposure to such an environment.

(3) Employment of a standard and expedient flotation devices in a water survival situation.

(4) Removal of individual field equipment after unscheduled water entry. Entry will be made from a minimum height of 10 feet with a complete set of field equipment, properly packed as an expedient flotation device and waterproofed. The ALICE pack will then be removed, the rifle laid across the top of the pack, the helmet remaining or put back on the individual's head, and the individual will traverse 25 yards to the side of the pool or water's edge.

c. The additional water survival qualification (flight physiology training) and requalification required of personnel assigned to flight status will be conducted in accordance with OPNAVINST 3710.7_,Promulgation of NATOPS General Flight and Operating Instructions. Swimming qualification (first-class, SI) is a prerequisite to water survival (flight physiology) training and is a one-time requirement.

Section II. Water Sports and Competitive Activities

4201.GENERAL

The commander, when developing the combat readiness program, needs to look carefully at the mission. If the mission requires Marines who must perform mission-essential water-related activities of long duration, he may want to look to water sports/competitive activities as a method of developing endurance. Endurance and confidence can only be developed by extended time in the water. While defined training can accomplish this goal, the goal can be enhanced through training which is integrated with the physical conditioning program. This approach will provide variety and enthusiasm. Two activities which will accomplish this goal are water relays and water polo. The important consideration is that the unit achieve total participation by all members of the unit.

4202.WATER RELAYS

Development of water relays will generally conform to the guidelines in chapter 5, section IV. When contemplating relays, the commander should review chapter 5 and adapt the principles to the size of the unit and the facility to be used.

4203.WATER POLO

The official rules provided in this paragraph will need to be adapted to the unit's unique situation, its size, and the size of the facility. Commanders should make every effort to have all members of the unit participate.

a. NCAA Rules This style of play is based on two popular sports: swimming and basketball. Playing is done on the surface of the water by teams of seven players each. The size of the playing area is 25 yards by 14 yards, which is the size of the typical high school or college indoor pool. The deeper the water, the better. If the entire playing area is deep-6 feet or more—this is ideal, although most indoor pools have a shallow end. At each end of the pool is a goal. In deep water, the goal is 10 feet across and 3 feet high. In shallow water, where the goalie can stand on and jump from the bottom, the goal is 10 feet across and 5 feet high. One of the seven players on each team is the goalie. He is the only player on the team who can stand on or jump from the bottom (if it is shallow enough to permit this), catch and pass the ball with both hands at the same time, or hit the ball with a clenched fist. However, the goalie may not swim across or pass the ball across the middle of the pool.

b.Equipment. The goalie on the visiting team must wear a white cap with the number 1 clearly marked on it; his teammates, who are designated as guards and forwards, must wear white caps numbered 2 through 7, with the substitutes wearing white caps numbered 8 through 21.The goalie on the home team must wear a dark-colored cap with the number 22 clearly marked on it; his teammates designated as guards and forwards, must wear similar dark-colored caps numbered 29 through 42.The guards and forwards on each team may swim freely up and down the pool as they see fit, interchanging positions as often as they wish. They may not stand on or jump from the bottom, touch the ball with more than one hand at a time when catching or passing or shooting it, or enter inside the opposing team's 2-yard line unless preceded by the ball. The ball is similar to a soccer ball, except that it is yellow and covered with a waterproofing substance that makes it easier to handle with one hand.

c. Playing Time. In intercollegiate competition the game consists of four 7-minule quarters, the teams changing ends after every period of play. In high school competition the game consists of four-,five-, or six-minute quarters, depending on the ages of the participants. Here again the teams change ends after every period. There is a very brief interval between each quarter. At the start of each period the teams line up at their respective ends, and with ablast of the whistle, the head referee throws the ball into the middle of the pool. The fastest swimmers on each team then race to gain possession.

d. Officials. The head referee walks along one side of the pool watching for infractions. He is aided by the assistant referee, who patrols the opposite side of the pool. The referees are both equipped with whistles and two official's flags. These flags should be 12 inches square and should correspond in color to the caps worn by the respective teams. Whenever a referee sees an infraction, he blows his whistle and signals with his flag indicating that the player nearest to the ball with the appropriate colored flag will be given possession of the ball. Whenever the whistle is blown, all players should react by looking immediately to see which color flag is being waived which will indicate which team should be moving into the defense. Other officials are a timer and a scorer. As in all sports, it is important that the officials be skilled and competent.

e. Technical Fouls. The following are some of the common technical fouls: starting before the referee blows his whistle to open play; holding onto or pushing off from the side of the pool during play; taking hold of the ball underwater when tackled by an opposing player; swimming inside an opponent's 2-yard line unless preceded by the ball; touching the ball with both hands at the same time (goalie excepted); standing, walking on or jumping from the bottom when taking an active part in the game. When a technical foul occurs, thereferee blows the whistle and with his flag awards possession. The player on the team awarded possession nearest the point of infraction then has 5 seconds to put the ball into play by passing to a teammate or by dropping the ball into the water and swimming after it.

f. Personal Fouls. The following are some of the more common personal fouls: committing any of the technical fouls mentioned above for the purpose of scoring or preventing a goal; holding, ducking, pulling, pushing off from, swimming over, or impeding the arm or leg movement of any opponent who is not touching the ball; and splashing water in the face of an opponent. When a personal foul occurs, the referee will follow the same procedure as with a technical foul. While doing this, the referee will also call out clearly and loudly the number of the player who was guilty of the infraction and a personal foul will be marked against that player by the official scorer. When a player accumulates five personal fouls, he fouls out of the game and must be replaced by a substitute.

g. Penalty Shot. A penalty shot can be awarded by either of the officials or the scorer when:

o An offensive player inside the opponents' 4-yard line not touching the ball is held, ducked, pulled-back, kicked, or struck.

o A team has accumulated a total of ten personal fouls. When the former occurs, the head or assistant referee should immediately blow the whistle and by holding a two-flagged stick in a vertical position above his head, signal that a penalty shot has been awarded. When the latter occurs, the scorer should use a buzzer located at the scorer's tabic to signal that one team has accumulated ten personal fouls, thereby entitling the other team to a penalty shot. A penalty shot is taken from the 4-yard line in front of the goal. All players except the defending goalie must leave the 4-yard line until the shot is taken. No player can be within 1 yard of the shooter. After ascertaining that the shooting player is on the 4-yard line and the goalie is on The goal line, the referee will ask the shooter to lift up the ball. When he docs, the referee will give a sharp blast from the whistle. At the whistle, the shooter must shoot without delay and without any faking at the goal. The goalie may try to block the shot. If he does, or the shot is otherwise missed, it is immediately in play and action continues.

h. When a Goal is Scored. When a goal is scored, either from a shot taken by a player out in the field or by penalty shot, the defending goalie must then pass the ball to the nearest referee. The referee will then pass the ball back to the goalie, who puts it into play by a pass to one of his teammates as soon as possible.

Section III. The Battle Swimming Test

4301.GENERAL

The battle fitness test is the evaluation contained in the Guidance for Basic Warrior Training (BWT) as part of the Basic Warrior Training Concept Plan.

4302.THE BATTLE SWIMMING TEST

The battle swim test consist of the following:

a. Be able to climb a three meter high board, while wearing a field uniform, carrying a M-16, web gear, and four canteens. Jump off blindfolded, recover and swim 100 meters with the M-16.

b. Repeat the same process, without the blindfold, and upon entering the water a second time, recover and retrieve all submerged gear.

c. Remain afloat and tread water for five minutes.

d. Demonstrate a knowledge of how to cross a stream using field expedient techniques and a single strand rope bridge.

e. Successfully complete helicopter egress training.

Chapter 5
COMPETITIVE CONDITIONING ACTIVITIES
Section I. Organization of Competitive Activities

5101.LEADERSHIP OF COMPETITIVE ACTIVITIES

a. Place in the Program. Competitive conditioning activities consist of dual combatives, relays, team contests, and team sports in which individuals or teams compete against an opponent to win. Competition is one of the best ways of maintaining interest in the physical training program during the sustaining stage. Organized competition provides enjoyable, vigorous physical activity that has proved to be one of the best supplements to conditioning drill activities. The benefits of competition are the development of aggressiveness, unit pride and identity, teamwork, and the will to win. These activities help develop cardiovascular and muscular endurance, strength, and coordination. Scheduling competitive activities in an orderly and progressive manner is desirable. The progression should be from relays to dual combatives, to team contests, and finally to team athletics. These activities should be conducted as part of the program after a basic level of conditioning has been developed, usually during the slow-improvement stage of conditioning. Muscles and joints should be strengthened by preconditioning to withstand the strain placed upon them by sudden stops and turns, body contact, bearing of weight, and falls. Competitive activities, however, should not be allowed to dominate the physical training program.

b. Area and Equipment. Some of the competitive activities included in this chapter require specific types of areas and equipment; others do not. The area requirement can usually be satisfied on available training fields. When items of equipment are required or specific courts or field layouts are to be marked off, such information is included.

c. Leadership. The principal factor for success as a leader of competitive activity is an energetic, dynamic, enthusiastic approach. The leader's attitude is reflected by the group, so the commander must carry on the activity in a snappy and vigorous manner. Confidence on the part of the commander will create an impression of decisiveness and certainty. Confidence grows out of experience and a thorough knowledge of the activity. Mastery of subject is the first step in developing confidence, assurance, and poise.

(1) The following suggestions are offered for leaders of competitive combatives, contests, and sports:

(a) Get the activity underway quickly by selecting and teaching only the essentials.

(b) Use rules to add to the enjoyment of the activity and do not allow them to interfere with the spirit of competition.

(c) Stop the activity before interest begins to lag.

(d) In team contests, clearly distinguish sides by attempting to maintain unit identity.

(e) Always insist on fair play, enforcing the rules impartially.

(2) The following procedure is recommended for presenting a competitive activity.

(a) Name the activity.

(b) Briefly explain the objective of the activity and give only the pertinent rules.
(c) Have a demonstration at slow speed and answer questions.
(d) Organize groups into teams and appoint captains.
(e) Arrange teams in the proper starting positions.
(f) Conduct the activity.

d. Competitive Units. Units for competition should be the same organizations in which Marines train: battalions, companies, platoons. In most situations, the unit is the squad.

e. Provisions for Instruction. One of the most effective methods of maintaining interest and participation in competitive activities is to provide instruction in activities with which most Marines are unacquainted. Such instruction can be conducted during regular physical training periods. Careful planning is required to keep all individuals continuously engaged in vigorous activity. The materials in the following sections should be used as a guide for instructional purposes.

f. Officiating and Control. Every effort should be made to provide good officials for all competitive activities. Poor officials quickly cause dissatisfaction among participants in team activities and create a situation which can turn an organized activity into a brawl. Each company should have or develop several competent and qualified officials available for games on company and platoon levels.

5102.MILITARY FIELD MEETS

When units reach the latter part of the slow-improvement and sustaining stage of training, interest in the program may lag. A change in course content can arouse the desire to participate. An event which does not require a high degree of skill yet demands strenuous activity is ideal for the military field meet. This is a series of team contests conducted on a station-to-station basis during a given period of time. Team contests carried on simultaneously provide essential training, vigorous exercise, and stimulating competition in an atmosphere that is enjoyable for all. Because of the healthy rivalry that a field meet arouses, it is an excellent form of inter-unit competition.

a. Objective and Advantages. The objective of the military field meet is to provide activity for everyone in the participating units. Activities are chosen that will develop aggressiveness, teamwork, a will to win, and competitive spirit, and that will stimulate interest and build esprit de corps. The military field meet can **be** included as part of the physical training program or as part of the off-duty recreational program. It is a form of contest that can be conducted in nearly all circumstances because it can be easily modified, requires a minimum amount of equipment, and can be readily organized.

b. Level of Competition. The military field meet is flexible. It can be adjusted to large or small groups. A company-size unit is the most desirable, but it may also be administered within a larger unit. If it is conducted within a company, the participating units will be squads. If it is held within a larger unit, platoons compose the teams. A larger-size unit requires more extensive organization and administration.

c. Selection of Events. In organizing a military field meet, select events that are simple and easy to administer. All rules and regulations should be clearly understood by everyone. No event should require previous practice. In selecting the events, consider the interest and capabilities of the Marines and available equipment and facilities. Select events which will require that all members of the units participate. Events should not be dominated by the athletes; rather, MOS-related events which have significance in producing skills and combat readiness should be used.

d.Equipment and Facilities. The site must be large enough to permit events to be grouped about a central control point. Use available facilities such as parade decks, Softball fields, or tracks. A public address system is desirable at the control point for the initial orientation of teams and for subsequent announcements of time lapses, cumulative scores, and final standings. A tally board is necessary so that all units can constantly monitor how the events are progressing.

e.Administrators. Efficient administration of the military field meet depends on the referees, judges, and scorers. For these positions, choose individuals within the units who have had athletic or officiating experience. Prior to the day of the event, all administrators should be briefed and assigned a specific task in order that they may become familiar with the rules and organization of the contests they will conduct. The following officials should be available:

(1) A primary instructor or supervisor who is in charge of the control center and who is responsible for the successful operation of the athletic carnival. The supervisor must have an assistant to act as a timer and scorer.

(2) One assistant instructor in charge of each event. This person should be-

(a) Familiar with the rules of the event.

(b) Effective as a leader to ensure proper supervision and control over the participating teams.

(c) Enthusiastic to provide proper motivation.

(d) Self-confident of the ability to judge infractions of the rules. The assistant instructor must be fair in judgment and penalize without hesitation when infractions occur.

(3) Runners between event stations and the control point. These individuals collect and deliver scores.

f. Team Organization for Competition. The size of the teams is determined by the level on which the field meet is organized. Maintaining the integrity of the unit promotes esprit de corps, but this does not preclude grouping two squads into one team.

g.Conduct of Events

(1) The assistant instructor at each station takes charge of the group and gives a brief explanation of the major rules of the event. The assistant instructor speaks clearly and distinctly from a position to be seen and heard by all. A short demonstration is desirable if it will help clarify the event.

(2) The assistant instructor should make certain that teams can be clearly distinguished by the use of, for example, T-shirts and fatigue jackets, or caps and no cap, or colored jerseys. The assistant instructor should teach a whistle response (teams stop play immediately upon hearing whistle), get the event started as quickly as possible, and make any necessary corrections as the contest progresses. The rale of good officiating is to use a minimum of calls, yet maintain control of the contest. Penalize when necessary, but refrain from disqualifying contestants or teams.

(3) Keep the activity moving as rapidly as possible. When the central control point sounds the whistle to stop the play, all competition ends immediately. The assistant instructor then assembles the group, forwards the team scores to the central point, and, upon a signal from the central control point, rotates the teams to their next station. It is essential that rotation and orientation of teams be quick and orderly to reduce time spent between contests.

(4) Upon completion of the final event, assistant instructors move their teams to the central control point for the announcement of winning teams, presentation of awards (if any), and final critique.

h. Scoring System. The system for determining the winner of the military field meet should be simple and efficient. At the completion of competition, the scorer totals the points that each team has scored in all contests. The scorer then subtracts the number of points scored against a team (penalty points for nonparticipation,etc.) from the total. The resulting scores are placed in a column with the highest score at the top and the lowest at the bottom. The team with the highest total is the winner. This type of scoring system encourages a team to ensure total unit participation.

Section II.Combatives

5201.DESCRIPTTON AND OBJECTIVE

Combatives are strenuous, short competitive contests in which two individuals attempt to overcome each other in a bout of skill and strength. These contests help to develop Marines' resourcefulness, confidence, strength, agility, coordination, and will to win. Any level ground area can be used. However, extremely hard ground should be avoided as some of the combatives require ground contact. A whistle is needed to control the bouts during the competition since voice commands may go unheeded. The extended rectangular formation is used for dual combatives. Combatives are conducted on an informal basis. Marines are allowed to remain at ease between activities and are allowed to brush themselves off after being on the ground. To get the most out of combatives, individuals must be urged to overcome their opponents as quickly as possible as would be required in combat.

a. Benefits for Marine Training. The possibility of close contact with an enemy in combat faces Marines at all times. They must be trained to react aggressively and violently. Combatives may be used as an introduction to such hand-to-hand contact and should be followed by hand-to-hand combat training. Marines enjoy competition, and this type of activity is a welcome change from formal conditioning activities.

b. Instructor Responsibility

(1) The instructor tells the individuals that all combatives begin and end on the whistle signal. The instructor demonstrates each activity before having it performed, explaining it in simple terms.

(2) After stopping one activity, the instructor gets everyone in place for the next.

(3) The instructor must closely supervise combatives to insure that contestants do not use unfair or unsportsmanlike tactics. To avoid unnecessary injury, instructors must see that the bouts are closely controlled and opponents equally paired. Adjustments should be made in apparent cases of mismatched abilities.

5202.COMBATIVES TABLES

There are three tables of combatives, each of which can be completed in 15 minutes. These tables become progressively more difficult from lower to higher numbers.

a.Combatives Table I

(1) Open-Hand Slap Boxing(See fig. 5-1, A. individuals assume a boxer's stance, palms open, fingers extended and joined. Each contestant tries to slap the opponent about the head and upper body with the open hand. This is a good warm- up activity.

(2) Wrist Tug-of-War(See fig. 5-1, B.)Two individuals sit on the ground with the soles of their feet in contact. Each grasps the opponent's wrists so that the hands are directly over the feet. At the whistle, each individual tries to pull the opponent from sitting to standing position.

(3) Arm-Lock Wrestling.(See fig. 5-1, C.)Two individuals sit back-to-back with legs spread and arms locked at the elbows. Each right arm is inside the opponent's left. Each tries to force the opponent's left arm or shoulder to the ground. The individual who first wins three bouts is the winner.

(4) Bulling(See fig. 5-1, D.)Two individuals assume the Westmoreland wrestling hold, each grasping the opponent's neck with the right hand and the opponent's right elbow with the left hand. Each tries to force the other to move one foot by pushing, pulling, or otherwise manipulating. The individual who first wins two boutsis the winner.

(5) Indian Wrestling(See fig. 5-1, E.)Two individuals lie on the ground, side by side, with their heads in opposite directions. They link right elbows. On the instructor's signal or by mutual agreement, each individual raises the right leg approximately straight and far enough to engage the opponent's heel. To start the contest, each person usually raises the leg three times rhythmically and on the third time engages the opponent's heel. Each tries to roll the other over backward. The right leg is used for three bouts, then the left leg for three bouts.

Figure 5-1 is available in the illustration section at the end of the book.

b. Combatives Table II

(1) Wrist Bending.(See fig. 5-2, A.) Opponents pair off, face each other, raise their arms forward, and, with palms forward, interlock their fingers. At the starting signal, each individual attempts to bend the opponent's wrist. The arms are kept up and forward and are not allowed to swing around and down to the sides. The individual who first wins two bouts is the winner.

(2) Back-to-Back Push.(See fig. 5-2, B.)Two individuals stand back-to-back with elbows locked. Each right arm is inside the opponent's left arm. At the starting signal, each pushes backward, trying to move the opponent. Opponents are not allowed to lift and carry each other; only pushing is permitted. The one who pushes the opponent the farthest wins the bout. The individual who first wins two bouts is the winner.

(3) Hop and Pull Hand.(See fig. 5-2, C.) The individuals are matched in pairs. Each grasps the opponent's right hand and, hopping on the right foot, attempts to pull the opponent off balance.Contestants automatically lose if they touch their free hand or their lifted foot to the ground. For successive bouts, they alternate hands and feet.

(4) Westmoreland Wrestling. (See fig. 5-2, D.)Each contestant grasps the back of the opponent's neck with the right hand and the opponent's right elbow with the left hand. In this position, each attempts to pull, push, or force the opponent to touch the ground with any part of the body other than the feet. The individual who first wins two bouts is the winner.

(5) Crab Fight.(See fig. 5-2, E.)Two individuals sit on the ground facing in opposite directions with their hands on the ground behind them. At the whistle, they raise their hips and push with their shoulders and bodies, each trying to make the other's hips touch the ground. The individual who first wins two bouts is the winner.

Figure 5-2 is available in the illustration section at the end of the book.

c. Combatives Table III

(1) Hand Wrestling.(See fig. 5-3, A.) Two individuals stand facing each other. Their right feet are forward and braced side by side. They grasp right hands for the first bout, left hands for the second bout. Each pulls, pushes, makes sideward movements, and otherwise maneuvers to force the opponent to move one or both feet from the original position. The contestant who first wins two bouts is the winner.

(2) Back-to-Back Tug.(See fig, 5-3, B.) Two individuals stand back-to-back with both arms linked at the elbows. Each has the right arm inside the opponent's left arm. At the starting signal, each attempts to pull the opponent. Lifting and carrying are permitted. The contestants must maintain their original direction and keep their arms linked. After a predetermined time, the player pulled or carried the farthest is the loser.

(3) Wrestling to Lift off Feet.(See fig. 5-3, C.) Contestants face each other. Each places the right arm under the left arm of the opponent and around the body. The left arm is over the opponent's right shoulder. Each tries to lift the other off the ground. The individual who first wins two bouts is the winner.

(4) Arm Pull Between Legs.(See fig. 5-3, D.) Two individuals are paired off, back-to-back. Each bends forward and, extending the right arm between the legs, grasps the opponent's right wrist. At the starting signal, each person attempts to pull the opponent After a predetermined time, the player who has pulled the opponent the farthest is the winner of the bout. The one who first wins two bouts is the winner. Repeat with the left hand and then both hands.

(5) Rooster Fight.(See fig.5-3, E.) Each contestant grasps the left foot with the right hand from behind, and the right arm with left hand. Each hops on the right foot, and, by shoulder-butting the opponent or by feinting and sudden evasions, forces the opponent to let go of the foot or arm. The contestant who first wins two bouts is winner.

Figure 5-3 is available in the illustration section at the end of the book.

Section III .Relays

5301. DESCRIPTION AND OBJECTIVE

Relays are races in which each member of a team runs one leg of the race. The team effort decides the winner. Relays provide stimulating competition and contribute to the conditioning of personnel. They also develop aggressiveness, team spirit, and the will to win. Relays should be dispersed throughout the program for short periods of time to provide a change of activity.

a. Team Organization. Relays are conducted most efficiently in platoon-size groups. Teams of equal size must be organized. Competitive spirit is encouraged and teams are organized better by basing teams on units such as squads, crews, or sections. Team captains should be designated. Extra Marines may be used as officials. The number of individuals on a relay team should be limited to squad size. If larger teams are used, the runners will spend too much time awaiting their turns and too little time actually participating. Two to six teams are ideal for relay competition. It is difficult to keep track of winners when more teams compete.

b. Administration of Relays

(1) The time spent on any one relay should be relatively short. If one team achieves a substantial lead in a long relay, the competitive spirit and enthusiastic participation of the other teams may decrease. Several short relays are generally better than one long relay.

(2) To maintain competitive spirit throughout a number of relays, determine the teams that win, place, and show in each relay and their total points for all relays. This can be done by awarding points to all teams on the basis of position at the finish of each relay. The team with the greatest number of points is the winner of the entire set of relays.

(3) Difficulties commonly encountered in conducting relays may be avoided by the following procedures:

(a) The last player in a relay race should be conspicuously identified. For example, the last runner can tie a handkerchief around the head or arm, take off the shirt, put on a hat or take it off, or use some other means.

(b) Another way to keep track of the progress of the race is to have each player sit or squat as soon as each is finished.

(c) Judges at the starting line can keep the runners from starting too soon.

(d) To prevent contestants from turning before they run the full distance, they should be required to run around a peg, pole, or assistant instructor.

(e) Batons, handkerchiefs, tent pegs, or other objects should be passed from one runner to the next when relays are run on a circular track.

(4) Before a unit's first participation in a relay, inform participants of the rules and scoring system. Violation of the rules should not result in disqualification. Instead, impose point penalties. A point penalty is imposed by subtracting a point from the team total at the conclusion of the relay.

(5) Careful administration will prevent most violations. For efficient conduct of relays, follow this procedure:

(a) Announce the name of the relay.

(b) Form the individuals in relay position.

(c) Briefly explain thc rclay and thc rulcs for running it.
(d) Demonstrate.
(e) Have a definite finish line, and ensure that all know where it is.
(f) Answer questions.
(g) Conduct the relay.
(h) Determine winner and award points.

5302. RELAY TABLES

a. Events. The following relays are grouped into a table of activities. Each relay table can be completed in 15 minutes. Thus, relays can be used as a sole activity or as a part of a longer period. Each table is planned for a platoon-size group (30-60Marines). Adequate warmup for participants is provided by conducting several repetitions of exercises 1 and 2 of a conditioning drill. The recommended relay tables require an area 40 by 60 meters in size. Each table provides a variety of activity. The tables are progressive in the overload applied and should be scheduled in numerical order although not necessarily on successive days.

b. Relay Table I
(1) **60-Meter Lane Relay** (See fig. 5-4, A.) Each team is assembled in single file behind the starting line. On signal, the first individual of each team runs to the turn-around line 30 meters away, then runs back and touches the next teammate waiting at the starting line. The winning team is the first team to get its last member across the finish line. If an individual starts before being touched by the preceding runner, the team may be penalized.
(2) Wheelbarrow Race.(See fig. 5-4, B.) The players of each team pair off and line up in single file behind the starting line. The first individual walks on the hands while the partner grasps the ankles. They advance to the distance line (25 meters) behind which they exchange positions and return to the starting line. The rear individual must always hold the partner's ankles. After the first pair returns across the starting line, the next pair starts.

Figure 5-4 is available in the illustration section at the end of the book.

(3) Squad Front Relay .(See fig. 5-4, C.) The teams form in a line along the starting line with a 10-foot interval between teams. The members of each team lock elbows so that they are linked together. At the starting signal, the teams run to the distance line (20meters) where the left flank individual acts as a pivot. The team swings around on the pivot and returns to the base line. If a team breaks its links, it must reform before continuing. The first team to completely cross the base line intact is the winner.
(4) Crab-Walk Race.(Scc fig. 5-4, D.) The players of each team line up in single file. The first person of each team assumes the crab-walk position with feet forward on the starting line, hands on the ground behind it. At the starting signal, contestants move forward to the distance line (10 meters). They touch the line with their feet and then return to the starting line in the reverse position with the head and hands leading. The second person may not start until the first individual touches the finish line.

c.Relay Table II

(1) 100-Meter Lane Relay.(See fig. 5-5, A.)This relay is conducted exactly as the first relay of table 1 except that the start and turn-around line are 50 meters apart. This relay provides progression in sprinting.

(2) Frog-Jump Relay. (See fig. 5-5, B.) Each team lines up in single file. The first individual assumes a squatting position on the starting line. At the starting signal, the contestant progresses to the distance line (15 meters) and back by leaping forward, catching the weight on the hands, and bringing up the legs to the squat position for the next leap forward.

(3) Simple Relay. (See fig. 5-5, C.)Each team lines up in single file. Place a marker on the distance line (20 meters) in front of each team. Each team member, behind the first one, grasps the belt of the individual ahead. At the starting signal, each team runs as a unit to the marker, circles it, and returns to the starting line. The first team to completely cross the starting line intact is thewinncr.

(4) Fireman's Carry Relay.(See fig. 5-5, D.)The players of each team line up in pairs, one behind the other. One individual in each pair carries the partner to the distance line (30 meters), using the fireman's carry. At the distance line, individuals exchange places and return to the starting line. As a variation, the person to be carried lies on the ground. The one carrying lifts the partner to the proper position. This relay may be performed with the other carries described in guerrilla exercises in chapter 3.

d.Relay Table III

(1) 200-Meter Circle Relay .(See fig. 5-6, A.)A course is laid out in a circular, rectangular,or oval pattern that is 200 meters around. Each team provides one runner on the starting line. On signal, the runner races around the 200-meter track and touches the next teammate, waiting at the starting line, who runs the same course. Each team member runs one lap of the course.

Figure 5-5 is available in the illustration section at the end of the book.

(2) Bear-and-Crab Race.(See fig. 5-6, B.) Each team lines up in single file. At the signal to start, the first contestant in each column assumes the bear-walk position (hands on ground in front), walks to the distance line (15 meters), and then runs back to the starting line. The returning contestant touches off the second individual and goes to the rear of the line. The second contestant assumes the crab-walk position (hands on ground behind) and crab-walks with feet leading to the distance line. This contestant also runs back to the starting line and touches off the next individual, who walks bear fashion. The rest of the members of each team alternate in this manner. The relay ends when the first person is back at the head of the line.

(3) Pilot Relay.(See fig. 5-6, C.) The players are grouped in threes, arms interlocked at the elbows, and outside players facing backwards. The middle person runs forward; the two outside individuals run backward. They run to the turning point (15meters), where they start back, this time with the middle person running backward and the two outside individuals running forward. The next set of three players starts when the first set crosses the starting line.

(4) Saddle-Back Relay.(See fig. 5-6, D.) Mark two parallel lines 15 meters apart. Each team selects a rider. The remaining members of each team count off. The even-numbered players from each team form in single files behind one line and the odd-numbered players from each team form in single files behind the other line directly across from their teammates. At the starting signal, the rider mounts the back of the first player of the team who carries the rider across the other line. There the rider changes mounts to the first player in the second line without touching the ground. This person carries the rider to the next player waiting in the first line. The relay continues until all of the mounts have carried the rider. A rider who falls off must mount again at the point of the fall. A rider who falls in changing mounts must get back on the original mount before making the change.

e. Relay Table IV

(1) 100-Meter Circle Relay (See fig. 5-7, A.)A course is laid out in either a circular, rectangular, or oval pattern that is 200 meters around. Each team is divided in half with each half positioned at starting lines on opposite sides of the track. Each runner races halfway around the track and touches a teammate who completes the lap. Each runner then waits in file at the halfway point until touched by the runner and then completes the second half of the lap. The first team to return all runners to their original starting line is the winner.

Figure 5-6 is available in the illustration section at the end of the book.

(2) In-and-Out Relay.(See fig. 5-7, B.) Each team lines up in a file with players 2 meters apart. At the starting signal, the first player runs back through the column in a zigzag fashion. He alternates, going to the right of one teammate and to the left of the next. Upon completing the run, the player lines up 2 meters behind the last one in the column. As soon as the first runner has passed the second person, the latter starts to run. This continues until all the players have realigned their original order. The team that finishes first is the winner. It may be desirable to have this relay continue until all individuals have run through their entire team two or three times in succession.

(3) Circle Race.(See fig. 5-7, **C.)** Each team forms a circle and holds hands with all individuals facing out except one who faces in and is the "driver."At the starting signal, the teams race to the distance line (20 meters) and back, keeping the circle intact. All the individuals in the circle must completely cross the distance line. The "driver" gives directions and orders. When the circle breaks, it must be reformed before it can continue. The first team completely over the starting line is the winner.

(4) Horse-and-Rider Relay.(See fig. 5-7, D.)Each team lines up in a single file. At the signal to start, the second player in each column leaps upon the back of the first who carries the rider across the distance line (30 meters). At the distance line, the rider dismounts and runs back to the starting line. There the rider picks up the third individual in the column, and carries this player to the distance line where the first player has remained. This continues until the last person is carried across the distance line.

Figure 5-7 is available in the illustration section at the end of the book.

Section IV. Team Contests and Athletics

5401. DESCRIPTION AND OBJECTIVE

Team contests are competitive activities in which Marines as a team compete with another team to win. They are guided by simple rules and organization. The function of team contests is to provide competition and an opportunity for body contact and to contribute to the development of physical readiness. In competing and working together as a team, individuals develop aggressiveness, the will to win, and teamwork. Team athletics deserve a prominent place in the physical training program because they contribute to increased combat efficiency. Because of the competitive nature of athletics and their natural appeal, individuals take part with enthusiasm. Athletic teams formed at intramural and higher levels are a strong unifying influence and provide one of the best means of developing esprit de corps.

a. Area and Equipment A level training field is sometimes the only area required. Many contests need no equipment. In contests requiring equipment, the need is for standard items such as logs, balls, nets, goals, and similar types of equipment. Specific requirements for area and equipment are listed with each contest description.

b. Progression. Team contests of a strenuous nature should be introduced after a basic period of conditioning has been completed and individuals are in the slow improvement stage of conditioning. Progression can take place from the less active to the more vigorous contests, and then from the non-contact to the contact or combative-type contests.

c. Necessity for Preconditioning Individuals must undergo conditioning prior to participation in athletics. Muscles, organs, joints, and ligaments not accustomed to stress and strain from sudden stops and starts, falls, body contact, rapid turns, prolonged running, and other rigors of athletic competition are subject to injury. Although athletics should not be introduced until players are physically prepared, there is still opportunity to engage in competition through lead-up contests. Individuals learn many of the skills required for athletics while participating in team contests.

d. Benefits. Athletics are beneficial primarily to sustain interest in the program and to maintain an achieved level of physical fitness. Athletics are a supplement, not a substitute for other types of conditioning activities which should continue. All of the desirable traits of physical fitness cannot be developed through athletics, yet their contribution is significant. For athletics to make a proper contribution to physical conditioning, the selected sports must bevigorous. The team contests recommended here are carefully selected for their simplicity, aggressiveness and applicability to squad- teaching teamwork. They are generally strenuous and involve the whole team throughout the entire duration of the contest.

5402. TEAM CONTESTS

a. Pushball

(1) **Players.** 10 to 50 players on a side. (See fig. 5-8.)

(2) Equipment A large pushball 5 to 6 feet in diamcter.
(3) Area. A field 240 to 300 feet long, 120 to 150 feet wide. Mark a center line in the middle of the field parallel to the end lines. Mark a line 45 feet on either side of this center line and parallel to it, extending it across the width of the field. Mark another parallel line 15 feet from each end line, extending it across the width of the field.
(4) Game. Four 10-minute quarters are played. Give 2-minute rests between quarters and 5-minute rests between halves. The object of the game is to propel the ball over the opponent's goal line by pushing, rolling, passing, carrying, or any other way except kicking.

(a) The ball is placed on the center line with the opposing captains 3 feet from the ball. The rest of the players are 45 feet from the ball, on their half of the field. On the referee's starting whistle, the captains immediately play the ball with their respective teams coming to their assistance.

(b) At quarter time, the ball remains dead for 2 minutes at the spot where it was when the quarter ended. At half time, the teams exchange goals. The play is then started as it was in the beginning.

Figure 5-8 is available in the illustration section at the end of the book.

(c) Players may use any means of interfering with an opponent's progress except striking and clipping. Clipping is throwing the body across the back of an opponent's leg, while the opponent is running or standing. Legal use of force may be applied to all opponents whether or not they are playing the ball. For striking an opponent, the offender is removed from the game. The team penalty is half the distance to their goal. The penalty for clipping is the same.

(d) When any part of the ball goes out of bounds, it is a dead ball. The teams line up at right angles to the side lines and 3 feet apart at the point where the ball went out. The referee then tosses the ball between the teams.

(e) When, for any reason, the ball remains in one spot for more than 10 seconds, the referee declares the ball dead. The ball is then put into play as it is for an out-of-bounds situation.
(5) Scoring. A goal is scored when the ball, or any part of it, is propelled across the opponent's end line. A goal counts 5 points. The team scoring a goal has the privilege of trying for a point after the goal. To try for this extra point, the ball is placed on the opponent's 5-yard (or 15-foot) line. The teams line up on either side of this line separated by the width of the ball. Before the whistle blows, one player on each team may place hands on the ball. On the referee's signal, the ball is put into play for 1 minute. If any part of the ball is driven across the goal line in this 1-minute period, the offensive team scores 1 point. The defending team may not score during the opponent's try for the extra point.

b. Line Rush
(1) Players. Any number up to 50 on each side. (See fig. 5-9.)
(2) Equipment. None.
(3) Area. A field, 75 by 100 feet.
(4) Game. One team lines up behind one goal line and the other in midfield. On the starting signal, the team standing behind the goal line seeks to cross the goal line at the opposite end of

the field within 30 seconds. The team in the center seeks to prevent it by catching and holding the runners. At the end of 30 seconds, the teams change.

(5) Scoring. Count the number of individuals who have crossed the far goal at the end of 30 seconds. After each team has had from three to five tries, the scores are added and the winner declared. A player scores 1 point when any part of the body is across the goal line.

Figure 5-9 is available in the illustration section at the end of the book.

c. Human Tug-of-War

(1) Players. 10 to 20 on a team.(See fig. 5-10.)

(2) Equipment. None.

(3) Area.40 to 60 feet.

(4) Game. Draw a line in the center of the area. Divide the players into two equal teams. Place them in single file on opposite sides of the center line facing each other. Each individual places the arms around the waist of the teammate in front. The two leaders of the opposing teams grasp each other around the waist. On signal, the teams try to pull each other over the center line within 30 seconds.

(5) Scoring. The team pulled across the center line loses. If neither team is pulled over the center line, but one team breaks its file, that team loses the match.

(6) Variation. Use a 3/4- or 1-inch rope. Space the leading players on each team 10 feet apart, each holding one end of the rope. The team pulled across the center line loses.

Figure 5-10 is available in the illustration section at the end of the book.

d. Master of the Ring

(1) Players. Any number. (See fig. 5-11.)

(2) Equipment. None.

(3) Area. A clearly marked circle large enough to contain all the players.

(4) Game. All the players stand inside the circle. At the signal, all players attempt to throw each other out of the circle. All tactics are fair except unnecessary roughness. When any part of the body touches across the line, the player is out and leaves the circle at once. Several officials are needed to spot the players who cross the line.

(5) Scoring The player who remains in the circle when all the others are out is the master of the ring.

(6) Variation l. The players are divided into two equal teams. Each team is clearly marked. On signal, each team tries to throw the opponents out of the circle. The winning team is the team that eliminates all opponents from the circle.

(7) Variation 2. The players are divided into two equal teams. Each team sends only one individual into the circle. When one person has been forced out of the circle, the losing side sends in only one player. The team which eliminates all opponents is the winner.

(8) Variation 3.A pit, approximately 4-feet deep, is used rather than a circle on level ground. The contest may then be an individual or team activity.

Figure 5-11 is available in the illustration section at the end of the book.

e. Log Pivot Circle. Space teams far enough apart so each team can pivot in a circle without colliding with any other team. Each team holds a log in the bend of the arms in front of the chest. (See fig.5-12.) At the command **CIRCLE RIGHT, MOVE**,the left-flank individual holds the pivot while the log is carried around 360 degrees in a counter- clockwise motion, back to the original position. This movement may also be performed to the left by facing in the other direction and pivoting clockwise (**CIRCLE RIGHT, MOVE**).Other commands may be used such as: **CIRCLE HALF-RIGHT, CIRCLE HALF- LEFT** and so on. The first team to complete the prescribed movement is the winner.

Figure 5-12 is available in the illustration section at the end of the book.

f. Log-Rolling Race. Each team tries to roll its log a measured distance by pushing it with the hands and driving the body forward with the legs. (See fig. 5-13.)The first team to get the entire length of the log across the finish line wins.

Figure 5-13 is available in the illustration section at the end of the book.

g. Prone Pushing Contest.. Two teams lie prone, facing each other with a log between them. (See fig. 5-14.)Both teams place their hands against the log, keeping their arms straight. Then, by driving with the legs, each team attempts to push the other a measured distance to the rear.

Figure 5-14 is available in the illustration section at the end of the book.

h. Shuttle Relay Race. Each team in this race divides into two parts, A and B. Members of part A run 50 yards with a log held under their right arms. (See fig 5-15.)At the distance line, they give the log to their teammates in part B who bring it back to the starting line. The team pair finishing first is the winner.

Figure 5-15 is available in the illustration section at the end of the book.

5403.CROSS-COUNTRY AND DISTANCE RUNNING

a. Value of Running. Long-distance running gives some benefits that cannot be obtained in the same degree from any other sport. It builds powerful leg muscles, increases lung capacity, and develops endurance. For these reasons, cross-country and distance running should be included in the physical training program. These sports require only a few miles of open space which is generally available. Short cross-country runs and middle-distance runs can be used to supplement other activities, particularly team sports or the sports that develop precision or agility rather than endurance. Short cross-country runs can be scheduled once a week, gradually increasing the distance as the physical condition of the runners improves.

b. Cross-Country Runs. A cross-country run is a distance run held on a course laid out along roads, across fields, over hills, through woods, and on any irregular ground. A flat cinder or dirt track is not a suitable surface for cross-country running. Some runs are as long as 6 miles while others are as short as 3 or 4 miles. If cross-country running is being used to supplement other activities, the 3-mile course is long enough for most. Cross-country runs should be used only after individuals reach the sustaining stage of conditioning. This type of running should then be scheduled occasionally to provide variety. Cross-country running has the advantage of allowing mass participation. Interest can be stimulated by putting the runs on a competitive basis.

c. Practice Methods. Conditioning is essential to distance and cross-country running. Championship distance running depends on stamina, which can be developed only through constant training. An individual of only average ability can become an outstanding distance runner by steady and careful training. Hiking is an excellent method for getting into condition before the season opens. Long walks build up leg muscles. During the first month of the season, training should be gradual, starting with short distances, and increasing day by day. At first the legs will become stiff, but the stiffness gradually disappears if running is practiced for a while every day. To prevent strain, it is essential to limber up thoroughly each day before running.

(1) In training a large group, leaders should be stationed at the head and the rear of the column. They should make every effort to keep the runners together. After determining the abilities of the unit in cross-country running, it is advisable to divide them into three groups. The poorest-conditioned group is started first, the best-conditioned group, last. The starting time of the groups should be staggered so that all of them come in at about the same time.

(2) In preliminary training, running is similar to ordinary road work in that it begins with rather slow jogging, alternating with walking. The speed and distance of the run is gradually increased. As conditioning improves, occasional sprints may be introduced. Atfirst, the distance run is from one-half to 1 mile. It gradually increases to 2 or 3 miles. On completing runs, require runners to walk for 3 or 4 minutes before stopping to permit a gradual cooling off.

d. Facilities and Equipment. There should be at least one timer with a stopwatch (preferably three) for timing the runners. A course 3 or 5 miles long should be measured and marked by one of the three following methods:

(1) Fasten directional arrows to the top of tall posts and place them at every point where the course turns. Such signs should also be placed at every other point where there may be doubt as to the direction of travel.

(2) Place a lime line on the ground over the entire course.

(3) Place flags, clearly visible to the runners.

- **(a)** A red flag indicates a left turn.
- **(b)** A white flag indicates a right turn.
- **(c)** A blue flag indicates the course is straight ahead.

e. Rules/Scoring. A cross-country team consists of seven individuals, unless otherwise agreed. In dual meets, a maximum of 12 on one team may be entered, but 7 or fewer are scored. First place scores 1 point, second place 2, third place 3, and so on. All who finish the course are ranked and tallied in this manner. The team score is then determined by totaling the points scored by the first five of each team to finish. The team scoring the least number of points is the winner. If fewer than five (or the number determined prior to the race) finish, the places of all members of that team are disregarded. If two or more teams score the same number of points, the event is called a tie.

5404.TEAM ATHLETICS

Team athletics add variety to the combat readiness program. At no time should they be allowed to replace intramural activities or to provide practice time for intramural or base teams. If team athletics are integrated into the program, the goal should be total participation and competition. Therefore, the games selected can be played by the skilled or the unskilled. They require limited equipment, yet will teach competitiveness, aggressiveness, and teamwork while they improve conditioning.

a.Soccer. Soccer is one of the best athletic activities for developing endurance, agility, leg strength, and great skill fulness in using the legs. The most popular sport in the world, soccer is the national game of many European, Asian, and Central and South American countries. In recent years, it has become popular in the United States. A soccer ball is the only equipment needed for the game, and Marines can learn to play it easily. Players do not need much skill to participate, but the amount of skill they can develop is almost unlimited.
(1) Place in the Program. Soccer should be introduced during the latter part of the slow improvement stage of physical training. It can be used as a competitive activity in the sustaining stage. It is primarily a spring or fall sport. Any level field is suitable for competition. The boundaries for the soccer field are similar to the dimensions for a football field. (See fig. 5-16.) Goal posts are essential, but they are easily constructed and usually temporary so that they may be removed when not in use.

Figure 5-16 is available in the illustration section at the end of the book.

(2) Basic Skills

(a) Passing. Passing with the feet is the primary means of moving the ball. Short passes are easier to control. Emphasis should be continually placed on the skill of passing.

(b) Dribbling. The ball is dribbled by a series of kicks with the inside or outside of the foot. Do not kick with the toe. Keep the head over the ball when kicking and propel it only a short distance at a time. Keep it close to the feet. If a player lets a ball move far from the feet, an opposing player can easily take it away.

(c) Instep Kicking. The instep kick is the basic soccer kick. The toe does not come in contact with the ball. The toe is pointed downward, and the instep (the area under the shoe laces) is applied to the ball with a vigorous snap from the knee.

(d) Inside-of-the-Foot Kicking. The ball is kicked with the inside of the foot, and the leg is swung from the hip. The toe is turned outward, and the sole of the foot is parallel with the ground as the foot strikes the ball. This kick is used for short passes and for dribbling.

(e) Foot Trapping. The foot trap is the method of stopping the ball by trapping it between the ground and the foot. Place the sole of the foot on top of the ball at the instant it touches the ground, keeping the foot relaxed. This is an effective way to stop a fast-moving ball.

(f) Shin Trapping. .The shin trap is a method of stopping the ball with the shins. Stand just forward of the spot where the ball should strike the ground and allow it to strike the shins in flight or on the bounce. Use either one or both legs from the knee down so as to absorb the bounce.

(g) Body Trapping. .Intercept the ball with any part of the upper body except the arms and hands. Keep the body relaxed and inclined toward the ball. To keep the ball from bouncing, move backwards from it as it strikes the body. This drops the ball at the feet in position for dribbling or passing.

(h) Heading. Heading is a technique for changing the direction of the flight of a ball by butting it with the head. Tense the neck muscles and jump up to meet the ball. Butt the ball with the forehead at about the hairline to reverse its direction. Use the side of the head to deflect it to the side.

(3) Offensive and Defensive Positions. Forwards usually play on the offensive half of the field. Fullbacks usually play on the defensive half of the field. Halfbacks are the backbone of the team. They move forward on the offense and back on defense. The goalkeeper almost always remains within a few feet of the goal.

(4) Abridged Rules

(a) A soccer team is composed of 11 players. (See fig. 5-16.)

(b) The player propels the ball by kicking it with the feet or any part of the legs, by butting it with the head, and by hitting it with any portion of the body except the arms or hands.

(c) The goalkeeper is the only individual allowed to place hands on the ball, but only in the goalkeeper's area. The term "hands" includes the whole arm from the point of the shoulder down.

(d) A goal is made by causing the ball to cross completely the section of the goal line lying between the uprights and under the cross bar.

(e) Each goal scores 1 point for the team scoring the goal.

(f) The penalty for a foul committed anywhere on the playing field (except by the defensive team in its penalty area) is a free kick awarded to the opposing team.

(g) All opponents must be at least 10 yards from the ball when a free kick is taken.

(h) The penalty for a foul committed by the defensive team in its penalty area is a penalty kick.

(i) A penalty kick is a free kick at the goal from the spot 12 yards directly in front of the goal. The only players allowed within the penalty area at the time of the kick are the kicker and the defending goalkeeper.

(j) An official game consists of two 30-minute halves,

(k) Teams change goals at the end of every quarter.

(l) In the event of a tie, an extra quarter may played. This may be followed by a sudden death period. If the tie still exists, the game can be decided by penalty kick.

(m) After a team has propelled the ball across a side line and out of play, the ball is put back into play by a throw-in from the side line by a member of the opposing team. The ball is thrown in from the point where it crossed the side line as it went out of bounds. The ball must be thrown in with both hands using an over-the-head motion, keeping both feet on the ground.

(n) When the offensive team propels the ball across the defensive team's goal line but not in the goal, the defensive team is awarded a goal kick-a free kick taken from within its own goal area. The ball must come out of the penalty area to be in play.

(o) When the defensive team causes the ball to go behind its own goal line, excluding the portion between the goal posts, the offensive team is awarded a corner kick—a free kick taken by a member of the offensive team at the quarter circle, at the cornerflag post nearest to where the ball went behind the goal line. The flag post must not be removed.

(p) The game is started and, after a goal has been scored, is resumed by placing the ball in the center of the midfield line. Players must be on their side of the line until the ball is kicked. The ball must be kicked forward and must move at least 2 feet to be legal. The first kicker may not touch the ball twice in succession at the kick off. The opposing team must be 10 yards from the ball until it moves.

b.Speedball. Speedball offers vigorous and varied action with plenty of scoring opportunities. It is easy to learn and provides spontaneous fun. Little equipment is needed-a ball is all that is absolutely necessary. Speedball combines the kicking, trapping, and intercepting elements of soccer; the passing game of basketball; and the punting, drop kicking, and scoring pass of football.

(1) Place in the Program.Speedball, like soccer, should be introduced into the physical training program during the latter part of the toughening stage and used as a competitive activity in the sustaining stage. Speedball may be played any time weather permits but is primarily a spring or fall activity.

(2) Method of Play.The game is played as follows:

(a) Two teams of 11 each play the game under official rules, but any number of players may successfully constitute a team.An inflated leather ball, usually a soccer ball, is used. The playing field is a football field with a football goal post at each end. (See fig. 5-17.)

(b) The game starts with a soccer-type kickoff. The kicking team tries to retain possession of the ball and advance it toward the opposite goal by passing or kicking it. Running with the ball is not allowed, with the result that there is no tackling or interference. After the ball touches the ground, it cannot be picked up with the hands or caught on the bounce but must be played as in soccer until it is raised into the air directly from a kick. Then the hands may be used.

(c) When the ball goes out of bounds over the sidelines, it is given to a player of the opposing team and is put into play with a basketball throw- in. When it goes over the end line without a score, it is given to a player of the opposing team who may either pass or kick it onto the field.

(d) When two opposing players are contesting possession of a held ball, the official tosses the ball up between them as in basketball.

(e) Points are scored by kicking the ball under the crossbar of the goal post, drop kicking the ball over the crossbar, or completing a forward pass into the end zone for a touchdown.

Figure 5-17 is available in the illustration section at the end of the book.

(3) Basic Skills

(a) Shared Skills. Skills include kicking, passing, heading, and trapping (from soccer); punting, dropkicking and forward passing (from football); and passing, receiving, and pivoting (from basketball).

(b) Kickups. The kickup is a play in which a player lifts the ball into the air with the feet so as to legally play the ball with the hands. The kickup generally makes the transition from ground play to aerial play. The technique of making the play depends upon whether the ball is rolling or stationary. If the ball is rolling or bouncing toward the player, the foot is held on the ground with the toe down until the ball rolls onto the foot. Then the foot is raised, projecting the ball upward. If the ball is stationary, the player rolls it backward with one foot. Then the player places a foot where the ball will roll onto it. The player then raises the foot, projecting the ball as before. If a ball is rolling away, the player should stop it with a foot and play it as a stationary ball. The player can also raise the ball by standing over it with a foot on either side. The player presses the feet against the ball and jumps into the air, propelling the ball into the hands.

(4) Offensive and Defensive Play. The positions of players in speedball are much the same as in soccer. However, some of the positions have different names. Each team has 11 players. The forward line is composed of five players, the right end, right forward, center, left forward, and left end. The second line consists of right halfback, fullback, and left halfback. In the next line is the right guard and left guard. The player who defends the goal is the goal guard. The strategy employed in speedball during offensive play is very similar to that of soccer. There are two types of defensive formations in speedball: man-for-man and position defense. Man-for-man defense is recommended for beginning players.

(5) Abridged Rules

(a) Field. A speedball field is 100 yards long with additional 10-yard end zones on each end and 55 yards wide, the same as a regulation football field.

(b) Players. Eleven players are on a team. The goal guard has no special privileges.

(c) Time. Ten-minute quarters. Two minutes between quarters. Ten minutes between halves. Five minutes for extra overtime periods.(Begin the first overtime by a jump ball at center, same goals; change goals in the event of a second overtime period.)

(d) Winner of Toss. The winner of the toss has the choice of kicking, receiving, or defending a specific goal.

(e) Kickoff. A kick off starts the first and second half of a game and starts play after a score. The kick off is made from any place on the 50-yard line. The kicking team must be behind the ball when it is kicked. The receiving team must stay back of its restraining line (10 yards distance) until the ball is kicked. The ball must go toward the opposing team's goal before it can be played. Kick off out of bounds goes to the opponent at that spot. A kick off touched by the receiving team before going out of bounds, with no impetus added, still belongs to the receiving team. A kick off, in possession and control of the receiving team and then fumbled out of bounds, belongs to the kicking team at that spot.

(f) Starting Second and Fourth Quarters. The ball is given to the team that had possession at the end of the previous quarter, from out of bounds, as in basketball.

(g) Second Half. The first-half receiving team kicks off at the beginning of the second half.

(h) Fly and Ground Ball. The most characteristic playing rule of speedball is the difference between a fly ball (or aerial ball) and a ground ball. A player is not permitted to touch a ground ball with the hands and must play it as in soccer. A fly ball is one that has risen into the air directly from the foot of a player (for example, a punt, drop kick, place kick, or kickup).Such a ball may be caught with the hands provided the catch is made before the ball strikes the ground again.

(i) Kickup and Overhead Dribble. A kickup is a ball that is so kicked by a player that he can catch it himself. A bounce from the ground may not be touched with the hand because it has touched the ground since being kicked. This rule prohibits the ordinary basketball dribble, but one overhead dribble (throwing the ball into the air and advancing to catch it before it hits the ground) is permitted.

(j) Out of Bounds. If a team causes the ball to go out of bounds over the sidelines, a free throw-in (any style) is given to the opposing team. When the ball goes over the end line without scoring, it is given to the opponents who may pass or kick from out of bounds at that point.

(k) Tie Ball. In case two players are contesting the possession of a held ball, even in the end zone, a tie ball is declared. The ball is tossed up between them.

(6) Scoring

(a) Field Goal(3 points).A soccer-type kick, in which a ground ball is kicked under the crossbar and between the goal post from the field of play or end zone, is a field goal. (A punt going straight through is not a field goal for it is not a ground ball. The ball must hit the ground first.) A drop kick from the field of play that goes under a crossbar does not count as a field goal. A drop kick from the end zone that goes under the crossbar counts as a field goal;if it goes over the crossbar, it is ruled as a touch back, and the other team takes possession of the ball.

(b) Dropkick(2 points). A scoring drop kick must be made from the field of play and go over the crossbar and between the uprights. The ball must hit the ground before it is kicked (usually with the instep).

(c) End goal(1 point). This is a ground ball which receives its impetus (kicked or legally propelled by the body) from any player, offensive or defensive, in the end zone and which passes over the end line but not between the goal posts.

(d) Penalty Kick(1 point).A ball kicked from the penalty mark that goes between the goal posts and under the crossbar is a penalty kick. The penalty mark is placed directly in front of the goal at the center of the goal line.

(e) Touchdown(1 point). A touchdown is a forward pass from the field of play completed in the end zone. The player must be entirely in the end zone. If the player is on the goal line or has one foot in the field of play and the other in the end zone, the ball is declared out of bounds. If a forward pass is missed, the ball continues in play, but it must be returned to the field of play before another forward pass or drop kick may be made.

(7) Substitutions. Substitutions may be made any time when the ball is not in play. A player who is withdrawn may not return during that same period.

(8) Timeout. Three legal timeouts of 2 minutes each are permitted each team during the game.

(9) Fouls

(a) Personal Foul. Kicking, tripping, charging, pushing, holding, blocking, and unnecessary roughness of any kind, such as running into an opponent from behind, are personal fouls. Kicking at a fly ball and thereby kicking an opponent is a personal foul. Four personal fouls disqualify a player from the game.

(b) Technical Foul. Illegal substitution, more than three timeouts in a game, unsportsmanlike conduct, and unnecessarily delaying the game are technical fouls.

(c) Violation. Traveling with the ball, touching a ground ball with the hands or arms, double overhead dribble, violating tie ball, and kicking or kneeing a fly ball before catching it are violations.

(d) Penalties. The offended player shall attempt the kick.

(e) Officiation of Fouls

1 Fouls in the field of play allow no follow-up while fouls in the end zone always allow follow-up.

2 On penalty kicks, with no follow-up, only the kicker and goalie are involved.

3 On penalty kicks, with a follow-up, the kicking side is behind the ball and the defending side behind the end line or in the field of play. No one is allowed in the end zone or between the goal posts except the goal guard. The kicker must make an actual attempt at goal and cannot play the ball again until after another player plays it.

c. Volleyball. Volleyball is a popular sport. The game entails much physical activity, yet it is not strenuous. It is, therefore, a game for young and old alike, for beginners and for skilled players. It may be played indoors or outdoors on any type of terrain. As an informal activity, volleyball can be played by any number; as an organized activity, it provides, as few other sports do, a game for 12 to play in a limited area. While volleyball requires no great skill to play, it permits a high degree of proficiency. Players naturally get more enjoyment when they know the game and play it well. For this reason, instruction in the basic skills should be provided.

(1) Place in the Program Volleyball may be used occasionally as a competitive activity during the sustaining stage. Though it is a year- round sport, it should be included in the physical training program only when a more strenuous activity is impractical. Volleyball is an excellent off-duty activity.

(2) Instruction. Usually during the first three or four classes, a 10- to 15- minute period of instruction, followed by scrimmage, is enough to teach the basic skills, rules, and techniques of volleyball. More time, if available, can be given to teaching basic skills, but emphasize competitive play rather than formal instruction. Divide the unit into 6-person teams, organized on the basis of ability .Teams should be as nearly equal as possible.

(3) Basic Skills

(a) Passing the High Ball. The chest pass is the most effective method of playing the ball. To receive the ball, the knees are fixed, and the body is tilted forward. The elbows are raised sideward to a point in line with the shoulders. The wrists are extended in line with the forearm. The arms, wrists, and hands are rotated inward. To pass the ball, the hands are chest-high, thumbs pointing inward. The fingers are flexed, forming a cup, allowing them to contact the ball. On contact with the ball, the wrists are snapped while the fingers and elbows are pushed upward, sending the ball upward. A high ball is much easier to handle than a low one.

(b) Passing the Low Ball. A ball that is lower than the waist is one of the easiest to hit, but it is also a frequent cause of fouls (holding or carrying the ball). The best position for handling a low ball is to have the knees flexed and arms flexed at the elbows and rotated so the thumbs are pointing outward, the palms up. When the fingers contact the ball, the entire body reacts in a lifting motion. The arms and hands swing upward in a scooping action. It is important that the fingers, not the palms, contact the ball, and that the ball is batted, not thrown.

(c) Underhand Serve. Take a position behind the back line facing the net, left foot forward, holding thc ball in thc palm of the left hand. The left knee is flexed; the right knee is straight. Swing the right arm back and at the same time move the left hand (holding the ball) across the body in line with the right hip. Then swing the right arm forward hitting the ball off of the left hand with the palm of the right hand, raising the hips and arching the back in the same motion. Be certain to swing the right arm in a straight line, or the ball will be difficult to control. When the opposition is in formation, the server should try to place the ball in the right or left back area, and not near the net.

(d) Setup. A setup is a ball hit into the air near the net by one player, so that a teammate may hit or "spike" it sharply downward into the opponent's court. The chest pass is the best pass to use. The ball is sent approximately 10 feet into the air toward the spiker so it will descend from 4 to 20 inches from the net.

(e) Spiking. The spike is a leap into the air and a sharp downward hitting of the ball into the opponent's court. A spiker must be able to spring easily from the floor, judge the movement of the ball, and strike it with a downward movement of the arm. To jump from the floor, step off with one foot and jump with the other. Stand with the right or left side to the net, facing the setup player. Much depends upon the setup player to place the ball in the proper position. The spiker jumps into the air and strikes the ball above its center so as to drive it downward. A snapping movement of the arm and wrist will drive the ball forward and downward with power and control. Aim for a weak spot in the opponent's defense.

(f) Blocking. The block is a defense technique used to prevent a spiker from driving the ball across the net. It is an attempt by one or more defensive players at the net to block a hard-hit shot by using the force of the ball to send it immediately back into the opponent's court. In an effective block, forwards on the defensive team spring into the air at the time of the spike, placing both hands and arms in the expected path of the ball. An effective block tends to upset the offense and presents another element for the spiker to worry about. To be effective, the blocker must anticipate the path of the ball and time the block with the spike.

(4) Abridged Rules

(a) The volleyball court is 30 feet wide by 60 feet long. (See fig. 5-18.)

(b) The top of the net is 8 feet high.

(c) A volleyball team consists of 6 players.

(d) A match consists of the best two out of three games.

(e) The first team scoring 15 points wins the game, provided that they have 2 points more than their opponents.

(f) A deuce game is a game in which both teams score 14 points. The game is continued until one team obtains a 2-point advantage over the other.

(g) Only the serving team can score. If the serving team commits a fault, it loses the serve to the opposing team.

(h) The team members who receive the served ball rotate one position in a clockwise direction.

(i) The ball is put into play by serving from behind the back line.

(j) A served ball touching the net results in the loss of the serve. At any other time during play, a ball touching the net is still in play.

Figure 5-18 is available in the illustration section at the end of the book.

(k) The ball is out of play when it touches the ground or goes outside one of the boundary lines.

(l) All line balls are good.

(m) The players must hit or bat the ball; they may not throw, lift, or grasp it.

(n) A player may not touch the ball with any part of the body below the knees.

(o) A player may not play the ball twice in succession. In receiving a hard-driven spike, a defensive player may make several contacts with the ball even if they are not simultaneous. All such contacts, however, must constitute one continuous play, and all must be above the knees.

(p) The ball may be touched no more than three times on one side of the net before being returned across the net to the opposing team.

(q) A player must not touch or reach across the net.

(r) A player may touch the line under the net, but may not cross it.

Chapter 6
EVALUATION OF PERFORMANCE DURING TRAINING

6001.OBJECTIVES OF COMBAT READINESS

The objectives of the Marine Corps physical fitness program are-

a. To contribute to the health and well-being of every Marine through regular exercise and health education.

b. To develop Marines who are physically capable of performing their duties in garrison and in combat.

c. To develop Marines with a reserve level of physical fitness that will help them win in combat.

d. To provide a medium for developing the self-confidence of the individual Marine and thereby enhancing overall discipline, morale, esprit de corps, unit efficiency, and the desire to excel within the Marine Corps.

6002. THE PHYSICAL FITNESS TEST

The physical fitness test (PFT) has distracted attention from the proper goal of the physical fitness program. Simply stated, the goal of the physical fitness program is the success of Marines in combat. Unit commanders must not allow the PFT to become an end in itself. The PFT should be used within the commander's program of combat readiness training as a standard, an index of the physical fitness of individual Marines at a given time and place. The PFT is not difficult, and it must be kept in perspective. All athletes should be first class, and first class is attainable for all Marines. The conduct of the PFT is described in Marine Corps Order 6100.3_.

6003. OTHER METHODS OF EVALUATION

The commander should use other ways of evaluation to provide variety and imagination. Intangible objectives, such as confidence and aggressiveness, are hard to measure. The use of either inspection or observation is necessary to evaluate objectives of this nature. The commander has several other methods available when evaluating the physical condition of the command. The additional methods of inspection, observation, medical examination, and testing are summarized as follows:

a. Formal Inspections. Formal inspections, using inspection officers and standardized rating criteria, may help evaluate unit physical fitness.

b. Daily Observation. Routine observation of physical performance and appearance can serve as an indicator of an individual's or unit's physical readiness. However, mere observation is not a totally reliable or accurate means of evaluation.

c. **Physical Examination.** Medical examination may detect any individual disability or detrimental physical condition. lt may guide in application of remedial, therapeutic, or limited exercise programs.

d. Other Physical Fitness Tests. Some other tests are described in detail in this manual for information purposes. Commanders are cautioned, however, not to place too great an emphasis on testing.

6004. RESPONSIBILITY

The commander is responsible for the physical fitness of the command and for the measurement and evaluation of its physical readiness. The goal of combat readiness training remains to prepare Marines and units for the physical demands of combat. The bulk of the commander's combat readiness training program should be such activities as obstacle and confidence courses, endurance courses, and progressive load-bearing marches which will prepare Marines for combat. Group activities which are competitive and combative in nature are advantageous and add imagination and enthusiasm to the program. Activities such as orienteering, which is not only physically demanding but also competitive and which teaches essential subject skills, are excellent and motivate all concerned. A commander's goal should be to provide an integrated training program that is so aggressive and demanding of Marines and leaders alike that a PFT administered at any time will reflect that each and every Marine is both physically and mentally prepared for the demands of combat.

6005. COMMANDER'S PHYSICAL FITNESS PROGRAM

The Commandant of the Marine Corps has directed that the physical fitness test shall be the universal measure of adequate individual physical fitness. The focus of a commander's physical fitness training program, however, should not be to prepare Marines to pass the physical fitness test, but to prepare them for the physical rigors of combat.

a. Administration. The physical fitness program requires each Marine to spend a minimum of 3 hours per week in physical fitness training, to be tested semiannually, and to obtain a minimum standard of third class. The attainment of a higher score is a laudable individual goal which should be encouraged but should not become a unit objective. Overemphasis could be detrimental to training required to develop the complete Marine. All commanders should exercise sound judgment as well as positive, aggressive leadership in striving toward the training objective.

b. Variety of Activities. The program must include the following elements:
(1) Physical Readiness Training. physical readiness training is a complete physical training program which develops and maintains the strength, endurance, and physical skills needed to sustain the individual during combat.
(2) Remedial Physical Conditioning. Remedial physical conditioning is a process by which physically substandard Marines are conditioned to meet prescribed standards.
(3) Competitive Conditioning Activities. Competitive conditioning activities consist of teams or individuals competing against an opponent to win. This involves a combination of sports and military skills designed to foster competitive attitudes and develop unit pride and esprit de corps.
(4) Occupational Conditioning. Marching long distances, dry net training, etc., should not be ignored as a means to fulfill the time requirement.

c.Time. In order to develop the desired level of physical fitness, unit commanders are advised that a minimum program of 3 hours per week is mandatory. The program must encompass at least three exercise periods weekly. Each period should include calisthenics, running, or other forms of vigorous activity .The minimum desirable length for each exercise period is 60 minutes. Shorter periods may be authorized by the unit commander if dictated by the local training situation. Unit commanders are encouraged to use the normal working day to satisfy this training requirement. However, commanders are authorized to conduct required physical fitness training during off-duty hours when the mission, workload, personnel status or other significant factors preclude fulfilling it during duty hours.

d.Action. Marine Corps Order 6100.3_ requires that commanders-

o Establish and maintain an effective physical fitness program which ensures that all Marines maintain an acceptable level of physical fitness.

o Ensure that all individuals are medically qualified prior to participating in the physical fitness program.

o Establish a minimum physical fitness program of at least 3 hours per week. Physical training may be authorized on an individual basis at the discretion of the unit commander.

o Ensure that all Marines participate in physical fitness conditioning activities commensurate with their medical qualifications and limitations.

o Conduct physical fitness testing for all Marines.

o Place all Marines who fail to pass the PFT on a daily command-supervised remedial physical conditioning program until they pass the test.

o Ensure that results of physical fitness testing are entered on section A of the fitness report. Comments are placed in section C in connection with attainment of superior physical performance or a medical excuse from the PFT.

6006. OTHER FORMAL TESTS

Commanders must realize that the best index of combat readiness is personal observation of performance on conditioning marches of 15 miles while under load. This manual presents three additional indicators of unit fitness. These indices are the physical readiness test, which was formerly the Marine Corps standard and is now routinely administered at Officer Candidate School; the airborne trainee physical fitness test which many Marines who attend U.S. Army formal schools are required to take; and the battle fitness test, the ultimate test of the combat readiness of Marines. Commanders may wish to include these tests or portions thereof within their combat readiness programs.

6007. PHYSICAL READINESS TEST

a. Purpose. The purpose of this test is to measure the Marine's ability to meet the minimum standard of physical readiness for duty in the field. This test is currently administered to all male officer candidates at Officer Candidate School, MCCDC, Quantico, VA. Female candidates do not take this test.

b. Test Area. Administration of this test requires an area suitable for crawling, jumping, and running 160 yards to include a trench that is 8 feet wide.

c. Equipment. The equipment required for this test includes—
o A platform, rail on a ledge, or a box 18 inches high,
o A 20-foot climbing rope,
o A stopwatch.

d. Conduct of Test. Adequate timers and other supervisory personnel must be provided by the company staff. All events should be conducted in a single session of one morning or afternoon. The events may be run in any sequence.

e. Uniform. The uniform for the physical readiness test is as follows: full utilities, cartridge belt, belt suspender straps, two canteens (full of water), rifle, and helmet. Depending on the heat condition, the PT shirt may be worn in place of the utility jacket. All equipment will remain secure throughout the test. Except for emergencies, canteens are not to be emptied.

f. Events

(1) Event 1: Climbing Uphill (Step-ups)This event simulates marching uphill at a rapid and steady pace. The Marine stands in front of the step. On the command GO, the Marine places either foot on the step and steps up. Hand pressure on the knee may be used if desired. The Marine then stands erect and steps backward and down, on one foot at a time. The same leg may be used for each repetition or the legs may be alternated. For 100 points, Marines will complete 60 repetitions in the maximum allotted time of 1 minute and 20 seconds (80 seconds).

*(2) Event 2: Rope Climb. T*his event resembles entering and leaving a hovering helicopter, using ropes in house-to-house fighting, and ascending and descending landing nets. The Marine assumes a sitting position on the ground at the bottom of the rope with hands at the highest point that can be reached. On the command GO, the Marine jumps up and climbs to the top. The Marine must touch the 20-foot mark with one hand, then descend without sliding or dropping. Marines must reach the 20-foot mark in the maximum allotted time of 30 seconds or less.

(3) Event 3: Evacuation. This event simulates reaching and evacuating a wounded Marine under fire. The Marine assumes the prone position at a distance of 50 yards from a casualty who is of approximately equal weight. On the command GO, the Marine springs up and covers the distance in a sprint. The Marine lifts the casualty to a fireman's carry and returns to the starting point. Assistance by the casualty is permitted. The tested individual must carry all equipment; the casualty has none. Marines must negotiate the course in a maximum time of 47 seconds or less.

(4) Event 4.'Advance by Fire and Maneuven. This event simulates advancing and assaulting during an attack. The Marine assumes the prone position at the starting point. On the command GO, the Marine then alternately crawls, runs in a zigzag manner, rolls, jumps, and assumes different firing positions for the next 130yards. Before reaching the objective, the Marine must "hit the deck," roll over, and assume the firing position three times. While assuming the firing position, the Marine must pull back the charging handle, take aim, and squeeze the trigger before rising from the ground. After rising from the ground for the third time,the Marine continues running and jumps to clear the 8-foot trench in one leap. The Marine then finishes the course in

the maximum allotted time of 2 minutes. The Marine is allowed multiple attempts only within the specified time limit.

(5) Event 5: Forced March. This event is a combat run of 3 miles without halts and against time. The Marine must double-time. Marines must complete the 3-mile course in a maximum allotted time of 30 minutes or less.

g. Event Standards

(1) To successfully pass the physical readiness test, the Marine must pass each event with a minimum of 80 points for each. Points are then assessed for each event in accordance with the scoring tables. (See fig. 6-1.)

(2) If a Marine fails one event, the highest possible score that the Marine may receive for the entire test is 79 percent. (See example 1 in fig. 6-2.) If two events are failed, the highest possible score for the entire test is 69 percent, regardless of the total points accumulated. (See example 2 in fig. 6-2.) If three events are failed, the highest possible score is 59 percent. Failure in four or more events will result in a score of zero for the entire test.

(3) To convert the total points for all five satisfactorily passed events to a final percentage score, total the accumulated points for all five events and divide by five. (See example 3 in fig. 6-2.)

Figure 6-1 is available in the illustration section at the end of the book.

Figure 6-2 is available in the illustration section at the end of the book.

6008. AIRBORNE TRAINEE PHYSICAL FITNESS TEST

a. Use and Composition of the Test.

(1) Use. The airborne trainee physical fitness test is a means of determining the physical ability of the applicant for acceptance to and retention in the airborne course of instruction.

(2) Test Events The test battery consists of five events as follows: chinups, knee bender, pushups, situps, and an endurance run.

b. Method of Scoring and Standards

(1) Scoring. The Marine will be scored by a trained scorer who is thoroughly familiar with the minimum standards for the test events. The Marine will be scored on a pass or fail basis. The performance on each event may be recorded on the reverse side of DA Form 705 (Physical Fitness Testing Record).The examinee records the personal information on the face of the card as specified on the first, second, and third lines. The face of the card is identified by writing diagonally across the lower half of the card "Airborne Trainee PPT—See reverse side."Using line 7 on the reverse side of the scorecard, the scorer enters the test title in the block entitled "(Other (Specify))" and completes the required information on the remainder of that line. The "Remarks" section can be used to record the test event titles and the applicant's performance on each test event.

(2) Standards. To successfully pass the test, the examinee must reach the standard in each test event. The standards follow:

(a) Chinups

(b) Knee bender 80(2-minute period)

(c) Pushups 45(2-minute period)

(d) Situps45(2-minute period)

(e) Endurance runs:2 miles in 15:54 minutes or less in athletic gear and 4 miles in 32 minutes in utilities and running shoes.

c. Uniform for Testing

(1) Examinees. The prescribed uniform for test participation is boots and the work uniform of the season. No headdress is worn. When climatic conditions permit, jackets or outer shirts may be removed.

(2) Officials. Scorers and other test officials should be uniformly and distinctively dressed for contrast with Marines being tested.

d. Test Administration Procedure

(1) Preparation for Test. The administration of the test to a large group makes it mandatory that the test effort be organized and efficiently operated. All testing is not completed with large test groups; on certain occasions, individuals and small groups are tested. Care must be exercised to administer the test uniformly and to standardize the conduct of all elements of the test. Regardless of the size of the test group, the following elements of sound test administration should be included:

(a) An orientation to include the purpose, method of administration, scoring of the test, preparation of the score card, and required standards.

(b) A correct demonstration of each event to ensure that there is no misunderstanding of the proper form and required standards.

(c) Completion of all five test events in one test period with all Marines taking the events in the same sequence.

(d) Adequate rest periods between the test events to allow for recovery before the next event.

(2) Method of Administration with a Large Group. With a minimum of 12 lanes per test event, 14 officials can administer the test battery to 150 or 200 Marines in 2 hours. The officials are designated as follows: one officer in charge, one demonstrator,and 12 scorers. If more or fewer Marines are to be tested, a greater or lesser number of officials will be required. The following procedure is recommended:

(a) Conduct an orientation and ensure the examinees have properly completed their score cards.

(b) Assign Marines to lanes and caution them to remain in the same lane order throughout the test.

(c) Explain and demonstrate the chinup event, administer it, and score it. Then proceed to the knee bender and pushup events, and administer them in the same manner.

(d) Grant a 5- to 10-minute rest period after the pushup event. Advise against excessive consumption of water during the break period.

(e) Explain and demonstrate the situp event, administer it, and score it as prescribed. Then move to the run area, explain the running event, and complete it.

(f) Retain the scorecards at the completion of the running event.

(3) Method of Administration with a Small Group. A similar procedure is followed for the testing of individuals and small groups. The informality usually associated with small groups must not conflict with sound test administration. With fewer examinees, fewer officials are required.

e. Description and Explanation of Test Events

(1) Test Event l:Chinups

(a) Purpose. This event is devised to test arm and shoulder flexor strength.

(b)Equipment There is one horizontal bar per lane, made of plumber's pipe or a gymnasium horizontal bar 1 1/2 inches in outside diameter. The bar must be rigidly supported at a height of 8 feet above the ground. The upright supports must be 5 feet apart. There must be a movable stand at each bar for short Marines to stand on to reach the bar.

(c) Officials There is one scorer per lane.

(d) Organization. The examinees, holding their scorecards, stand in order behind the restraining line in their respective lanes. The scorers take each scorecard when the Marine is called forward for the test.

(e) Starting Position The bar is grasped with the palms turned toward the face, the thumbs underneath the bar. The body is fully extended in a "dead" hanging position with the arms straight and the feet above the ground.

(f) Movement. Pull the body directly upward until the chin is placed over the bar. Lower the body until the elbows are completely straight and the body is again in the "dead" hanging position. Repeat as many times as required.

(g) Instructions. Explain and demonstrate the fully extended "dead" hanging position with the proper grasp. Show that the chin is placed over the bar at the top of the movement and that the arms are fully extended, the elbows completely straight, at the bottom of the movement (the hanging position). Explain that the body must be kept from swinging and that it is permissible to raise the legs and flex the hips when pulling up, but that any kicking, bicycling, or jerking motion with the trunk or legs is not acceptable. Inform the examinees that no penalty is exacted for hanging on the bar to rest in the bottom position but that this is not to their advantage. Tell them that half-completed chinups are not counted, and that the scorer will repeat the number of the last correct chinup when incorrect execution is detected.

(h) Administration and Scoring. Caution the examinees to assume the "dead" hanging position and wait for the scorer's command to begin. The scorer is at the examinee's left with a clear view of the bar. If the examinee begins to swing widely, the scorer should stop the swinging by extending the left arm across the front of the examinee's body, being sure not to hinder the execution of the chinup. The scorer counts aloud the number of chinups correctly executed. When a chinup is not correctly executed, the scorer repeats the number of the last correct one. The scorer records the number of correct chinups on the scorecard and returns the card to the examinee.

(2) Test Event 2:Knee Bender

(a) Purpose. This event measures the strength and endurance of the leg muscles.

(b) Equipment. None.

(c) Officials There is one scorer per lane.

(d) Organization. Marines stand in numerical order behind the restraining line in their respective lanes. The scorer takes each scorecard when the Marine is called forward for the test.

(e) Starting Position. The feet are spread less than shoulder width apart, hands on hips, thumbs in the small of the back, elbows back.

(f) Movement. Do a knee bend and at the same time bend slightly forward at the waist and thrust the arms between the legs until the extended fingers touch the ground. The hands are about 6 inches apart. The bend is approximately a three-quarters bend. From this knee bend position, recover to the starting position by moving the body upward, straightening the knees, and returning the hands to the hips. Repeat as many times as required.

(g) Instructions. Explain and demonstrate the correct starting position. Be certain examinees understand the correct knee bend and that only the tips of the fingers touch the ground. Tell them the scorer will repeat the number of the last correct knee bender when incorrect execution is detected. Some of the common errors are failure to correctly bend the knees, failure to touch the ground, and failure to assume the completely erect position after the bend has been executed.

(h) Administration and Scoring. The scorer stands to one side so as to see that the knees are properly bent and the fingers touch the ground as prescribed. From this position, the scorer can view the examinee to see that a properly erect position is assumed after each knee bend. The scorer counts aloud the number of correctly executed knee bends. When a knee bend is done incorrectly, the scorer repeats the number of the last correct one. The scorer records the number of correct knee bends on the scorecard and returns it to the examinee.

(3) Test Event 3:Pushups

(a) Purpose. Pushups measure arm and shoulder extensor strength.

(b) Equipment. None.

(c) Official. There is one scorer per lane.

(d) Organization. Marines stand behind the restraining line in their respective lanes until the scorer calls on them to perform. The scorer takes each scorecard when the Marine comes forward.

(e) Starting Position. The front-leaning rest position is the starting position. The body is straight from head to heels, palms are flat on the ground directly underneath the shoulders, and elbows are straight and locked. The body weight is supported on the hands and toes throughout the event.

(f) Movement. Bending only the elbows, lower the body in one straight plane until the chest touches the scorer's hand. Straightening and locking the elbows, raise the body in one straight plane, returning to the original front-leaning rest position. Repeat as many times as required, keeping the body in a straight line from head to heels.

(g) Instructions. Explain and demonstrate that the arms are straight at the beginning and completion of the movement and that the chest must touch the scorer's hand, but the stomach and thighs must not touch the ground. Also explain that the whole body must be maintained in a straight line as it is lowered and raised; that is, there is to be no breaking at the hips or shoulders so that any body part is lowered or raised in advance of the other or as a separate segment .Dipping or rolling through the shoulders is illegal, as is lowering or raising the body with one arm or shoulder at any time. Resting is not permitted during repetitions. Instruct

Marines that the scorer will repeat the number of the last correct pushup when incorrect execution is detected.

(h) Administration and Scoring. It is recommended that Marines assume a prone position while placing their feet and hands in the proper positions. This permits them to rest while the scorer gets into position and, at the same time, provides a feel of the body in a straight plane from head to heels. The scorer lies on the right hip and side to the right of the examinee. This gives the scorer a clear view of the examinee's body to see any errors. The palm of the scorer's right hand rests flat on the ground underneath the lowest part of the examinee's chest. By keeping the right forearm flat on the ground at an angle in front of the examinee's right arm, the scorer's position will not prevent the examinee from lowering the body completely. The scorer's left hand is free to test the straightening of the elbow at the completion of the movements and to point out body segments being lowered or raised separately. When in position and ready, the scorer has the examinee assume the starting position and begin doing pushups. The scorer counts aloud the repetitions done correctly and repeats the number of the last correct pushup if one is done incorrectly. There is no penalty if the contour of the examinee's body causes the hips to protrude slightly out of line, provided that the whole body is raised and lowered simultaneously. The scorer enters the number of repetitions on the scorecard and returns it to the examinee.

(4) Test Event 4: Situps

(a) Purpose. This event primarily measures abdominal strength.

(b) Equipment. None.

(c) Official. There is one scorer per lane.

(d) Organization Marines stand behind the restraining line in their respective lanes until the scorer calls on them to perform. The scorer takes each scorecard when the Marine comes forward.

(e) Starting Position. The examinee lies flat on the back with knees flexed, both feet flat on the ground. The correct angle of the thighs to the ground is 45 degrees. If the heels are too near the buttocks, the applicant will not be able to sit up. Legs are spread shoulder width apart. The examinee interlaces fingers and places them behind the head in contact with the ground. The feet are not held by another person.

(f) Movement. Bend forward at the waist and raise the upper body until the head is directly over the knees. Heels are not to leave the ground. Elbows remain in the same plane to the head and body throughout the event. The upper body is slowly lowered to the starting position until the head touches the ground. Repetitions are done at a slow cadence with no rest periods.

(g) Instructions. Explain and demonstrate the correct starting position and the proper execution of the situps to be sure that Marines understand the movement. Warn them that their knees must remain flexed during each situp, the heels cannot leave the ground at any time, and they may not roll up on one side and push up with one elbow. Tell them they must do the repetitions at a slow cadence with no rest periods. Instruct Marines that the scorer will repeat the number of the last correct situp when incorrect execution is detected.

(h) Administration and Scoring. When the performer is in position and ready, the scorer has the performer assume the starting position and begin doing situps. The scorer counts aloud the correct executions. When a situp is improperly done, the scorer repeats the number of the last correct one. No situp is credited if the hands are unclasped from behind the head, if the back is used to bounce up from the ground (which means the shoulders would not touch the ground), or if one shoulder or elbow is used to push up. The examinee is not penalized if the heels slide forward slightly so long as the knees remain flexed and the heels maintain contact with the ground. The scorer enters the number of repetitions on the scorecard and returns it to the examinee.

(5) Test Event 5:Endurance Run

(a) Purpose. This event measures cardiovascular endurance.

(b) Equipment One stopwatch or watch with a sweep second hand.

(c) Area. A large training field on which a quarter-mile track has been staked out or a level road over flat terrain may be used as a running surface. A 1-mile route is designated with wooden stakes marking the start point, finish point, and one-quarter mile intervals.

(d) Officials For large groups, a scorer times the event and controls the conduct of the run, and a guide runs with the group and sets the pace.

(e) Organization The run is conducted with groups of Marines in a column formation. Company-sized units may run at the same time with platoons serving as running groups. The scorer issues the command to assume double-time.

(f) Starting Position. Marines are assembled in proper column formation (column of twos, threes, or fours, as appropriate to the size of the group) with short men to the front. When all is ready, the column is moved forward a short distance before the running period is started.

(g) Movement. At the command DOUBLE-TIME, MARCH, the examinees retain their places in the column formation and execute the command. Length of steps is about 40 inches. The scorer has the group execute the run. The formation is maintained during therun.

(h) Instructions. Marines are instructed to maintain formation while running and are informed that the guide will set the proper pace. They will be instructed in the command used to control the column in the execution of the test. The scorer should announce the 4-minute, 2-minute, 1-minute, and 30-second remaining time intervals.

(i) Administration and Scoring. The event may be administered as previously prescribed to **a** large group, to several Marines, or to an individual. An individual examinee usually does not require a guide or pacer. If the event is administered on a training field, the scorer may stand in the center as the runner(s) circles about the field. This method of administration relieves the scorer of running with each group to be tested. Scoring is based on successful completion of the run as prescribed. The scorer should announce the remaining times as prescribed in (h).

6009. BATTLE FITNESS TEST

The battle fitness test is a physical fitness evaluation which is currently under development as part of the basic warrior training concept plan.

Chapter 7
THE HUMAN BODY

7001.GENERAL

To implement an effective physical conditioning program, a leader must understand how the human body functions and how exercise affects the body .This chapter provides only an introductory discussion of anatomy and functioning.

a. Body Functioning During the Stages of Conditioning. For more on the stages of conditioning, see chapter 1.

(1) During the toughening stage, the waste products of muscle activity (lactic acids) collect more rapidly than the blood can remove them. This acid waste builds up in the muscle tissue and irritates the nerves in the muscle fiber, causing pain and stiffness. As the exercise program continues, more blood is carried through the muscle, removing the waste materials more rapidly and eventually causing the soreness to disappear.

(2) During the slow improvement stage, the blood circulation in the muscles increases, and the body as a whole becomes a more efficient machine. The improvement is rapid in the first few weeks, but as a higher level of skill and conditioning is reached, the improvement becomes less noticeable. The body reaches its maximum level of performance after 6 to 10 weeks and should then be maintained at this peak.

(3) During the sustaining stage, it is possible to maintain this state of conditioning through 15 to 20 minutes of exercise a day, but the exercise must be quite strenuous.

b. Diet. Regular exercise increases the appetite. If the desire for greater amounts of food is satisfied by a balanced diet, the body benefits.

(1) There are two main types of foods: body-building and energy- producing.

(a) Body-building foods consist of proteins, which build up tissue and repair wear and tear.

(b) Energy-producing foods include carbohydrates and fats. Carbohydrates provide a quick source of energy, while fats act as a reserve store of energy.

(2) In addition, food contains vitamins, mineral salts, and water. During hot weather and strenuous training periods, the body requires greater fluid intake.

(3) Diet should be supplemented with proper rest to allow the digestive system to digest the lood.

(4) Occasionally, especially during early stages of conditioning, strenuous exercise may cause vomiting. Although vomiting is not a frequent occurrence, it is not usually a cause for concern.

7002. SYSTEMS OF THE BODY

The systems of the body include the skeletal, muscular, circulatory, respiratory, endocrine, digestive, genitourinary, and nervous systems. Each has a different function, but all must work in cooperation with one another to insure a sound body. Of these systems, the first four are the most affected by exercise and are discussed in succeeding paragraphs.

7003. THE SKELETON

a. Bones. The skeleton (see fig. 7-1) is composed of about 206 bones, which-

o Provide structure for the body and support for the attachment of muscles,

o Protect vital organs, such as the brain, lungs, and heart,

o Manufacture red blood cells, which carry oxygen through the body.

Figure 7-1 is available in the illustration section at the end of the book.

b. Joints. A joint is a place of union between two or more bones. Joints can be-

o Immovable, such as in the face and head.

o Slightly movable, such as in the vertebrae or sternum.

o Freely movable, such as in the shoulder, hip, knee, ankle, and elbow. The bones in a freely movable joint are held in place by ligaments. Freely movable joints are of the greatest importance in physical training because exercise improves their mobility and stability.

c. Cartilage and Ligaments. The joints in the body, where bones connect, are supported by cartilage and ligaments.

o Cartilage is a tough, elastic, translucent tissue that acts as a shock absorber or buffer between bones. Examples are the discs between the vertebrae, the tissue attaching the ribs to the sternum, and buffers in the knee joints. Damaged cartilage does not heal.

o Ligaments are connective tissue that bind bones together. They contract and extend but are not elastic. Consequently, ligaments can be torn or strained. Damaged ligaments will heal to some extent.

7004. THE MUSCLES

a. General. Muscles are tissue, or an organ composed of tissue, which contract or extend to produce movement. Muscles are of three classifications: involuntary, voluntary, and cardiac. The voluntary and cardiac muscles are the most affected by physical training; physical training has little effect on involuntary muscles.(Seefigs. 7-2, 7-3, 7-4.)

o People have no control over the involuntary muscles, an example of which is the diaphragm,

o Voluntary muscles are the larger skeletal muscles which cause movement and which the individual can control.

o Cardiac muscle is found only in the heart and for all practical purposes is an involuntary muscle. Physical training has a significant effect on the fitness of cardiac muscles.

b. Muscle Structure. Muscle is composed of cells which are grouped into bundles called fibers. Groups of fibers form muscles. These bundles of fibers are held in place by sarcolemma, a thin, sheath-like material that surrounds the muscle bundles and secretes a fluid that lubricates the muscle tissue. The fused ends of the sarcolemma form the tendons which attach the muscles to the bones.

c. Attachment of Muscles. The arrangement of muscles on the skeleton provides the proper angle of pull to make movement possible. Voluntary muscles are usually attached to the skeleton in two places.

d.Action of Muscles. To produce motion and to do work a muscle usually shortens its fibers. The movement may be flexion such as bending the arm at the elbow, or extension, such as the straightening the arm. Some muscles raise the arms or legs, others depress the raised limb. Some muscles have the primary function of rotating the trunk from side to side. In addition, muscles act as stabilizers as well as prime movers.

Figure 7-2 is available in the illustration section at the end of the book.

Figure 7-3 is available in the illustration section at the end of the book.

Figure 7-4 is available in the illustration section at the end of the book.

7005. FUNCTIONING OF THE SKELETON AND MUSCLES

a. General. The skeleton and muscles function in close coordination. Muscles move the body, and bones support the muscles.

b. Effect of Exercise on Bones. Continuous exercise, particularly among younger people, usually strengthens the bones, allowing them to withstand greater stress and strain. Bones which are not used lose much of their strength, a factor which should be considered when individuals return to physical training after a prolonged period of inactivity. People in this category should refrain from activities which might result in bone breakage before the bone is exercised back to normal condition. The condition known as "march fracture" is usually a result of this condition.

c. Muscular Strength. When exercised vigorously to improve strength, muscle grows in size. In general, the larger the muscle, the stronger the muscle. Furthermore, trained muscle functions more efficiently than untrained muscle. Trained muscles contract more vigorously and with less effort. Regular and strenuous exercise also toughens muscle, making it firmer and able to stand more strain.

d. **Muscular Endurance.** Muscles naturally become fatigued during continuous, repeated work. Through specific exercises, such as chinups or situps, local muscle groups become exhausted long before an individual fatigues. Training to develop muscular endurance enables people to continue a relatively heavy load of exercise over a long period of time. Lengthened exercise periods improve muscular endurance.

e. Muscular Coordination. Improved speed and strength result in part from improved muscular coordination. An unskilled individual may use irrelevant muscles to perform a particular activity, increasing the amount of work without increasing the mechanical output. This increase in skill is highly desirable, but it should be offset by greater effort in duration or intensity to compensate for the loss in overload due to increased skill.

f. Muscular Fatigue. When the rate of work is increased, the energy required is proportionately much greater than the increase in rate. For example, if an individual doubles running speed, the amount of power demanded to do this is increased eight times.

g. Circulation in Muscles Regular, strenuous exercise causes the creation of new capillaries and the opening of inactive, latent capillaries, increasing blood circulation within the muscle by as much as 400 percent. This increases the supply of food materials and oxygen to the muscle, improving its endurance. This process takes about 8 to 12 weeks of regular conditioning in young adults and longer as age increases.

7006. THE CIRCULATORY AND RESPIRATORY SYSTEMS

a. The Circulatory System. The circulatory system transports blood to all parts of the body, removes waste products for disposal, and delivers protecting and repairing substances where needed. The circulatory system includes the heart and blood vessels.

(1) The Heart. The heart is a pump which forces blood through the blood vessels to the parts of the body. The heart is a little larger than a fist and is located in the left center of the chest between the lungs.

(2) Blood Vessels. The vessels carrying blood away from the heart are the arteries. They eventually divide into capillaries, the very small vessels through which diffusion and osmosis take place. The capillaries gradually increase in size, forming the veins which carry blood back to the heart.

b. The Respiratory System. The respiratory system performs the function of breathing, whereby oxygen is inhaled and carbon dioxide is exhaled. The respiratory system consists of the mouth, nose, trachea, lungs, and diaphragm.

o Trachea. The trachea, or windpipe, is a hollow, tube-like structure that carries air from the mouth to the lungs.

o Lungs. The lungs are elastic bags in the chest where the exchange of oxygen and carbon dioxide takes place.

o Diaphragm. The diaphragm is a thin, sheet-like muscle just below the lungs. During inhalation, the diaphragm flattens out and lowers, allowing the lungs to expand and fill with air. During exhalation, the diaphragm raises into a dome shape, helping to expel air from the lungs.

7007. CARDIOVASCULAR FUNCTIONING

a. Gcncral. Cardiovascular functioning is the combined functioning of the circulatory and respiratory systems. The chief organs of these systems, the heart and lungs, function together to provide oxygen to the body.(See fig. 7-5.)

Figure 7-5 is available in the illustration section at the end of the book.

b. Heart Action. The heart, the organ which propels blood through the blood vessels, is the chief organ of cardiovascular endurance. A person tires quickly if the heart lacks the capacity to circulate the blood through the body. The heart is a muscular organ and is developed by exercise just like any other muscle. The heart cannot be exercised alone; any exercise that puts a sufficient load on the heart, such as running or cycling, will also exercise other body parts. A conditioned heart tends to beat more slowly but pump more blood when at rest. This is known as an increase in "stroke volume,"a desirable condition because it enables the heart to pump more blood with a slower contraction rate.

c. Functions of the Lungs. Strenuous and regular exercise can improve the efficiency with which the lungs transmit oxygen to the blood by as much as 25 percent In a poorly conditioned person, some of the alveoli (air sacs) within the lungs are closed or collapsed. The forced breathing created by exercise over a period of weeks causes the air sacs to slowly expand, increasing the ability to absorb oxygen.

d. Relationship of Heart and Lungs. The heart pumps carbon dioxide-laden blood through the pulmonary artery to the lungs. In the lungs the carbon dioxide is exchanged for oxygen and the purified blood is returned to the heart by way of the pulmonary vein. The heart then pumps the blood through the aorta for circulation throughout the body. Blood moving into the muscles exchanges oxygen for carbon dioxide, after which time it is ready for the return trip to the heart.

e. Cardiovascular Functioning in High Altitudes. Marines to be employed in high altitudes should be acclimated in a similar area for 10 to 14 days prior to employment. Persons not accustomed to the rarefied air of higher altitudes tire more quickly and may collapse after rapid physical exertion. Air is much less dense at high altitudes than at sea level. Persons accustomed to sea level inhale only about 80 percent of the oxygen they are used to no matter how hard they breathe. Further, people accustomed to sea level or moderate altitude simply do not have enough red corpuscles in the blood to fulfill their needs at high altitudes. The red blood cells will increase over time.

f. Symptoms of Cardiovascular Malfunctioning. The brain is the first organ to react to a lack of oxygen. Unconsciousness results when the brain is denied sufficient oxygen. "Blacking out" is actually a defense mechanism to keep the body alive.

APPENDIX A
INSTRUCTOR HINTS AND INSTRUCTOR TRAINING

1. General. This appendix informs instructors on administration of exercise periods, commands, the extended rectangular and circle formations, methods of instruction and conduct of exercise activities, explanation of basic positions associated with the various activities, and instructor training.

2. Aim of the Physical Fitness Program. The aim of the physical fitness program is to prepare Marines physically for war. This can be conveniently split into two phases.

a. Preparation Phase

(1) Development of all-around physical fitness.

(2) Development of purposeful physical skills.

(3) Development of mental alertness.

(4) Development of character and leadership.

b. Maintenance Phase. Once the preparation is completed, the program must progress to a maintenance program. This phase revolves around the need to maintain the trained Marine's physical fitness for a particular role. This will include—

(1) Maintenance of a high standard of all-around physical fitness.

(2) Application of purposeful physical skills to combat skills.

(3) Development of leadership and self-confidence.

(4) Development of mental toughness and endurance.

3. Physical and Mental Development

a. Marines need to be developed physically and mentally. The physical training (PT) instructor must ensure that the program accomplishes this.

b. The system of the body functions best in response to progressive exercise. This means that Marines should master the easier exercises before they attempt the more difficult ones.

c. The body and mind arc linked inseparably and must be considered as one in physical fitness training. Physical exercises assist in developing the mind.

d. The confidence in oneself and in one's comrades attained by physical achievement helps to promote the indomitable will to win. The Marine must be taught to realize this. It will help during arduous training, which at all times must be imaginative, realistic, and filled with enthusiasm. All physical fitness training must be associated with developing the qualities required of a Marine.

4. Planning and Preparation of ProgramsThe following factors must be considered when compiling a program (see also ch. 2)~

a. Aim or desired skill level.

b. Duration of the complete training program and number of physical training lessons.

c. Purpose of included exercises,drills, or practical applications.

d. Beginning/inventory fitness levels.

e. Medical facilities, medical personnel, and Marine Corps or local command regulations governing their proximity to the training.
f. Number of Marines to be trained.
g. Availability of instructors, assistants, and facilities.
h. Apparatus and equipment available.
i. Alternative training areas for varying weather conditions,
j. Stages of progression and tests of achievement.

5. Five Basic Principles of Program Design

a. Regularity. Regularity of exercise is far more important than the amount of exercise performed on a given day. A Marine should exercise a minimum of three times per week. Human muscle begins to atrophy after 3 days without exercise.

b. Balance. The program must be balanced in terms of total physical fitness. Imbalance is created when, for example, an individual merely runs. While the cardiovascular system is improved, general upper body strength may be lacking. Conversely, a person who just does weight training may develop strength, but not cardiovascular endurance.

c. Overload. If total body fitness is to be developed, then the muscular and cardiovascular systems must be made to perform more work than previously. This is done by increasing the frequency, intensity, and duration of the exercise. The body adapts to the amount of stress placed upon it and becomes stronger and more efficient.

d. Progression. To be successful, any type of training program must be progressive. This is an area where untrained persons conducting PT can cause injury to Marines resulting in failure of the PT program. Close supervision must be made of untrained instructors. The result of progressing too fast is that the class begins to reject PT because it is no longer enjoyable or motivating. The program should start at a slow pace, and the pace should increase as proficiency improves. Average persons will encounter three stages of progressiom the toughening stage, the slow improvement stage, and the sustaining stage. (See ch. 1.)

e. Variety. Some programs fail because the training becomes boring. Variety is essential and perhaps presents the greatest challenge to the commander and staff.

6. Grouping. Providing for different levels of physical fitness is particularly recommended in the early stages of conditioning. Older Marines and those in poorer physical condition should be expected to attain a group level of fitness, but they should be given more time to do it.

a. Homogeneous Grouping One simple method of providing for the difference in levels is to group them according to their condition. A two-group classification would divide Marines into highly conditioned and average groups. A finer classification could be obtained by dividing them into three groups—a highly conditioned, a moderately conditioned, and an unconditioned group.

b. Performance or Age Grouping. The segregation of Marines into different exercise groups should be based on physical fitness test scores or on the level of fitness they demonstrate. They may also be grouped at first according to age. A common classification by ages is under 30, 30 to 34, and 35 and above.

7. Preparatory Commands and Commands of Execution. The prepreparatory command describes and specifies what is required. The command of execution calls into action what has been prescribed. All preparatory commands are given with a rising inflection. The interval between commands is long enough to permit the average Marine to understand the first one before the second one is given.

8. Extended Rectangular Formation. The traditional formation for carrying on many physical training activities is the extended rectangular formation (fig. A-l). This formation is more compact than any other. It is the best type to use for large numbers of Marines because it is simple and easy to assume.

a. For the formation of oneplatoon, the base man paces off five paces from the stand, faces left and moves five paces, halts, and again faces left. With the base man positioned facing the stand, the platoon leader then commands. **FALL OUT AND FALL IN ON THE BASE MAN**. At this command, all Marines run to the designated area and reform. This procedure is preferred to marching the unit into position. If more control is desired, the unit may march at double time to the vicinity of the base man and then be directed to fall out and fall in on him. Much valuable time is wasted in the field by needless maneuvering of Marines at quick time in an effort to position the platoon or unit on the exact spot for the exercises.

b. A company-size unit assumes the extended rectangular formation from a column of threes or fours at normal intervals between squads. This extension can also be executed from a company mass without interval between platoons. In extending either a platoon or company-size unit, take your place at the head of the column and command:
(1) TAKE INTERVAL TO THE LEFT, MARCH. At this command, the Marines in the right flank file stand fast with arms extended sideward. All others turn to the left and run forward at double time. After taking a sufficient number of steps, all Marines face the front with both arms extended sideward. The distance between finger-tips is about 12 inches, and dress is right.

Figure A-1 is available in the illustration section at the end of the book.

(2) ARMS DOWNWARD, MOVE. At this command, the arms are lowered smartly to the sides.
(3) LEFT, FACE
(4) EXTEND TO THE LEFT, MARCH. At this command, those in the right flank file stand fast with arms extended sideward. All others turn to the left and run forward at double time. Spacing is the same as in (1) above, and dress is right.
(5) ARMS DOWNWARD, MOVE. Same as in (2) above.
(6) RIGHT, FACE.
(7) FROM FRONT TO REAR, COUNT OFF. At this command, the leading Marine in each column turns the head to the right rear, calls of **ONE** and faces the front. Successive Marines in each column call off in turn **TWO, THREE, FOUR, FIVE** in the same manner.
(8) EVEN NUMBERS TO THE LEFT, UNCOVER. At this command, each even-numbered Marine stride jumps sideward to the left, squarely in the center of the interval. In doing this, each one swings the left leg sideward, jumps from the right foot to the left foot, and smartly brings the right into position against the left.

c. To assemble the unit, you commandASSEMBLE TO THE RIGHT, MARCHAt this command, all return to their original position in the column at double time and reform on the base man.

d.It is recommended that the area for grounding equipment and arms be at the edge of, or well away from, the area to be used for exercising. To conserve time and ensure proper position of the unit, the base man or, if the unit is composed of several platoon-sizegroups, the various base men may precede the unit and establish their positions in relation to the instructor's stand.

9. Circle Formation.The circle formation is effective for conducting various exercise activities (fig. A-2). This formation has an advantage in that supervision of all Marines is facilitated and a moving formation is available which provides control. Guerrilla exercises, grass drills, and some forms of running are examples of activities which are more easily conducted in the circle formation than in the extended rectangular formation.

a. It is not advisable to have more than 60 Marines in a circle. When more must be accommodated, separate circles should be used.Concentric circles may be employed to reduce the size of the circle or to accommodate more persons. If concentric circles are formed, the different circles are made by designating squads for each circle.Each additional circle requires more Marines than the one inside it.For example, one squad of a platoon may form the inner circle and theremaimng three squads the outer circle.

b. When a platoon is to form a circle, the commands ai€IRCLEFORMATION, MARCH, FOLLOW MEUpon this command, the right flank squad of the column moves forward at double time with the leader of the platoon group gradually forming a circle in a counterclockwise direction. Each succeeding file falls in behind that on the right. After the rough outline of the circle is formed, the leader commandsPICK UP A 5-YARD INTERVAL.

Figure A-2 is available in the illustration section at the end of the book.

c.The group may be halted and faced toward the center, or, if instruction is not necessary, the exercise activity may be executed without stopping the platoon.

lO.Leadership Techniques

a. Unless you experience all the exercises, you cannot appreciate how arduous they are, what movements are most strenuous and difficult, where the errors in performance are likely to occur, and what the proper cadence should be.

b. You must give everyone careful supervision and participate in the exercises to show that you can do them. When you participate, your assistant instructors should supervise because it is difficult for you to supervise and exercise at the same time.

c. Marines should never be kept too long in one position, especially a constrained one. They should never have to perform an exercise more times than they can do it without losing the proper form. You should insist upon proper form in the execution of all exercise activities.Even slight deviations from the proper form reduce the value of the exercise.

d. Avoid long cxplanations.As a rule, it should be necessary to give a full explanation of new exercises only. Explain the most essential features of an exercise first; add details later. Too many details at one time arc more likely to confuse than to assist them.Minor corrections should be made to the entire class while the exercise is in progress (for exampHEADS **UP,** KNEES STRAIGHT). If necessary, follow this direction by the name of the person who is particularly at fault. If a Marine requires special attention, givethat person separate instruction by an assistant instructor to avoidwasting the time of the group.

e. The heavy demand on your voice can be lightened by training assistant instructors to assume some of the instruction. Using mass cadence is also an effective method of lessening the demand on your voice.

“Illustrations”

DAY	TIME	ACTIVITY
MONDAY	60 MIN	RUNNING ACTIVITY
TUESDAY	60 MIN	COMBATIVES
WEDNESDAY	60 MIN	CONDITIONING & LOG DRILLS
THURSDAY	60 MIN	CONDITIONING & RELAYS
FRIDAY	60 MIN	RUNNING & CONDITIONING
SATURDAY	60 MIN	RUNNING ACTIVITY

Figure 2-1. Sample A- -Single Activity.

DAY	TIME	ACTIVITY	REMARKS
MONDAY	20 MIN 20 MIN 20 MIN	RIFLE DRILL COMBATIVES RUNNING	PROGRESSIVE ACTIVITY SYSTEM
TUESDAY	30 MIN	GUERRILLA EXERCISES TABLE I	SINGLE ACTIVITY
WEDNESDAY	30 MIN	RIFLE DRILL	SINGLE ACTIVITY
THURSDAY	30 MIN	CIRCUIT INTERVAL TRAINING	SINGLE ACTIVITY
FRIDAY	20 MIN 20 MIN 20 MIN	STRENGTH CIRCUIT CONDITIONING DRILL CROSS-COUNTRY RUN	ROTATION ACTIVITY
SATURDAY	60 MIN	WARM UP RUN AND GRASS DRILL	SINGLE ACTIVITY

Figure 2-2. Sample B- -Various Time Periods, Activities, and Systems.

DAY	TIME	ACTIVITY
MONDAY	45 MIN	RUNNING – CROSS-COUNTRY
TUESDAY	75 MIN	TEAM ATHLETICS – TOUCH FOOTBALL
WEDNESDAY	45 MIN	RUNNING – CROSS-COUNTRY
THURSDAY	20 MIN 40 MIN	LOG DRILL DUAL COMBATIVES
FRIDAY	90 MIN	SPEED MARCH
SATURDAY	90 MIN	TEAM CONTESTS

Figure 2-3. Sample C--Sustaining Activities.

Route	Kilometers from SP	Recommended rate of march (kmph)	Remarks
SP: RJ 8th Div Rd—Superhighway	......	4	Bridge; hard surface; two guides.
RJ Jamestown—Hersey Rd	2.25	4	Traffic heavy; two guides; Jamestown Rd bears to right.
RJ Jamestown—Yankee Rd	7.10	4	Two guides
RJ Jamestown—Lightning Rd	10.50	4	Two guides
RJ Jamestown—Sunshine Rd	12.40	4	Light traffic; one guide Sunshine Rd; gravel, poor traction when wet.
RP; RJ Sunshine—Sedan Rd	14.80	4	Two guides

Figure 3-1.Example of Route Reconnaissance Report.

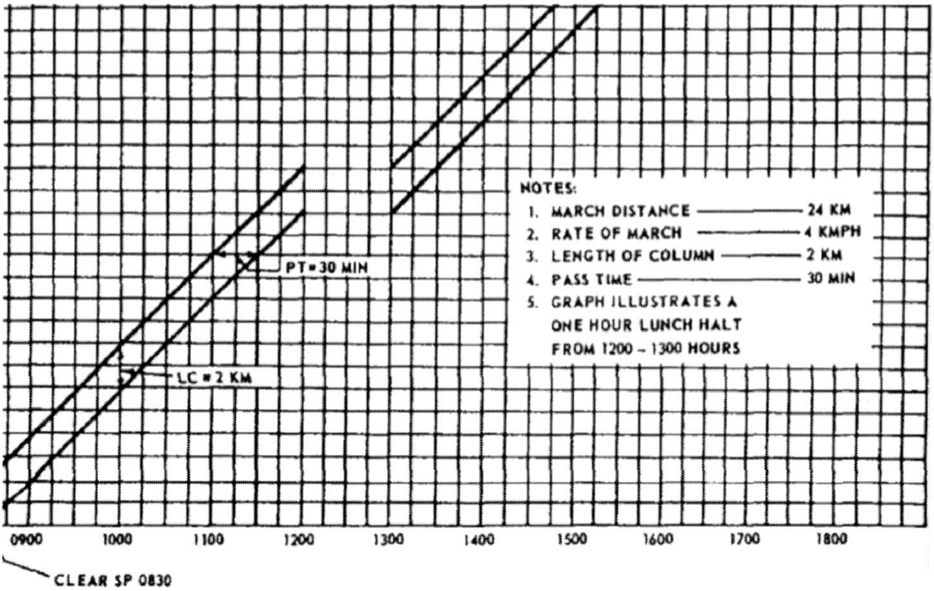

Figure 3-2.Road Movement Graph.

Figure 3-3.The Pace-Setter.

	SINGLE FILE	COLUMN OF TWO
2 m/MAN APART	24	12
5 m/MAN APART	54	27

GAP DISTANCES			
SERIAL:	(3-1) × 100 m	=	200 m
MARCH UNIT:	(12-1-2-) × 50 m	=	450 m
	TOTAL COLUMN GAP		650 m

Figure 3-4.Conditioning Drill One.

Figure 3-5.Conditioning Drill Two.

Figure 3-6.Conditioning Drill Three.

Figure 3-7. Rifle Drill.

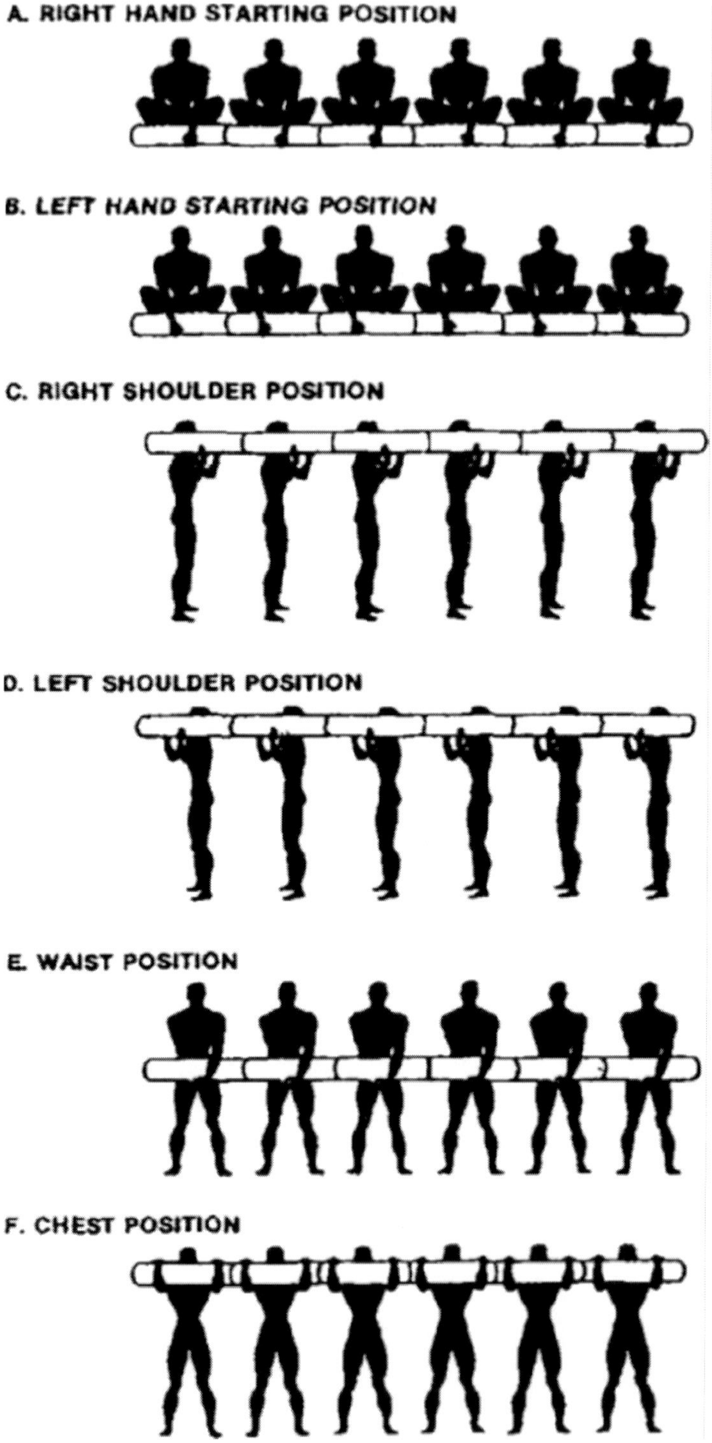

Figure 3-8. Starting Positions.

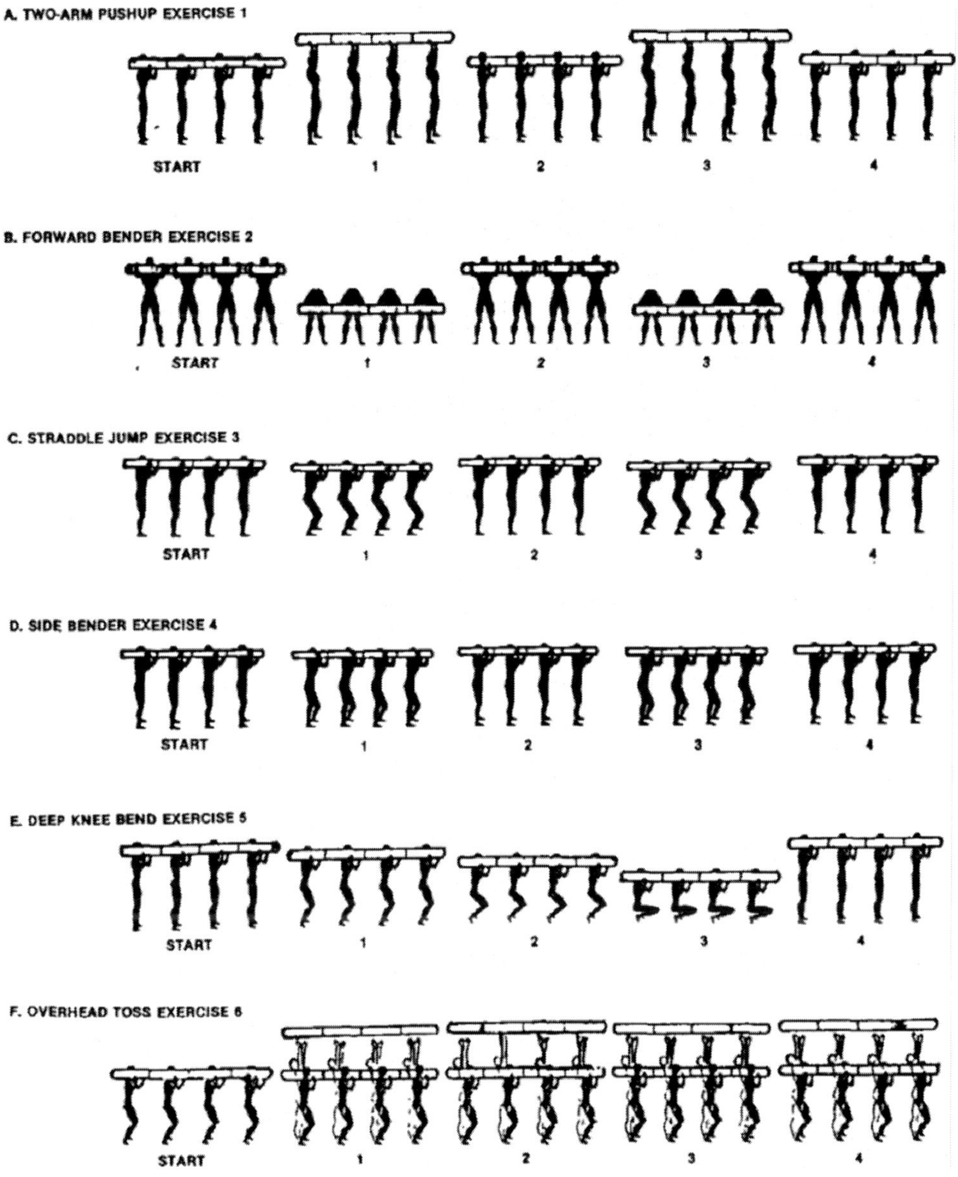

Figure 3-9. Log Drill

Figure 3-10.Basic Positions for Grass Drill.

Figure 3-11.Grass Drill One.

Figure 3-12.Grass Drill Two.

Figure 3-13.Guerrilla Table I.

Figure 3-14.Guerrilla Table II.

Figure 3-15.Proper Running Form.

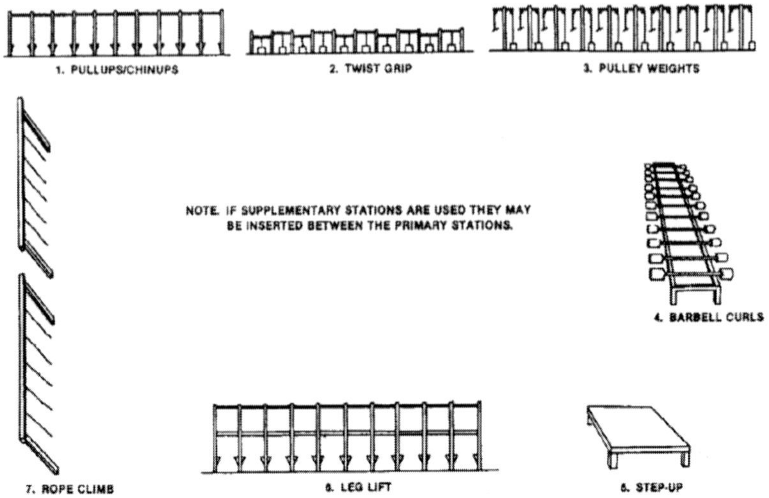

Figure 3-16. Fixed Strength Circuit.

STATION	ITEM	NO.	SPECIFICATION
1	BARBELL		1¼-INCH PIPE 5 FEET LONG WITH CONCRETE FILLED NO. 10 CANS
2	JUMP ROPE		¼- OR ⅜-INCH ROPE, 10 FEET LONG
3	TWIST GRIP		HANDLE 12 INCHES LONG, ROPE 4 FEET LONG, NO. 10 CAN CONCRETE FILLED.
4	INCLINE PLANE		¾-INCH PLYWOOD PLATFORM 2 FEET WIDE AND 6 FEET, 6 INCHES LONG, ELEVATED 10 INCHES AT ONE END. STRAP TO HOLD FEET DOWN.
5	WAR CLUB		HEAD IS 6 BY 12 INCHES, HANDLE IS 14 INCHES LONG BY 1¼ INCHES IN DIAMETER, ABOUT 20 POUNDS.
6	BICYCLE RIDE		PLYWOOD BOARD OR PLATFORM 2 BY 3 FEET WITH 2 BY 2 RUNNERS.
7	STEP—UP		A BOX OR STURDY PLATFORM 18 INCHES HIGH, 18 INCHES WIDE, 24 INCHES LONG.
8	ISOMETRIC PULL		TWO HANDLES 12 INCHES LONG WITH 4 FEET (BETWEEN HANDLES) OR LIGHT WIRE CABLE OR ¼-INCH ROPE

Figure 3-17.Movable Strength Circuit.

Figure. 3-18.Stations of Movable Strength Circuit.

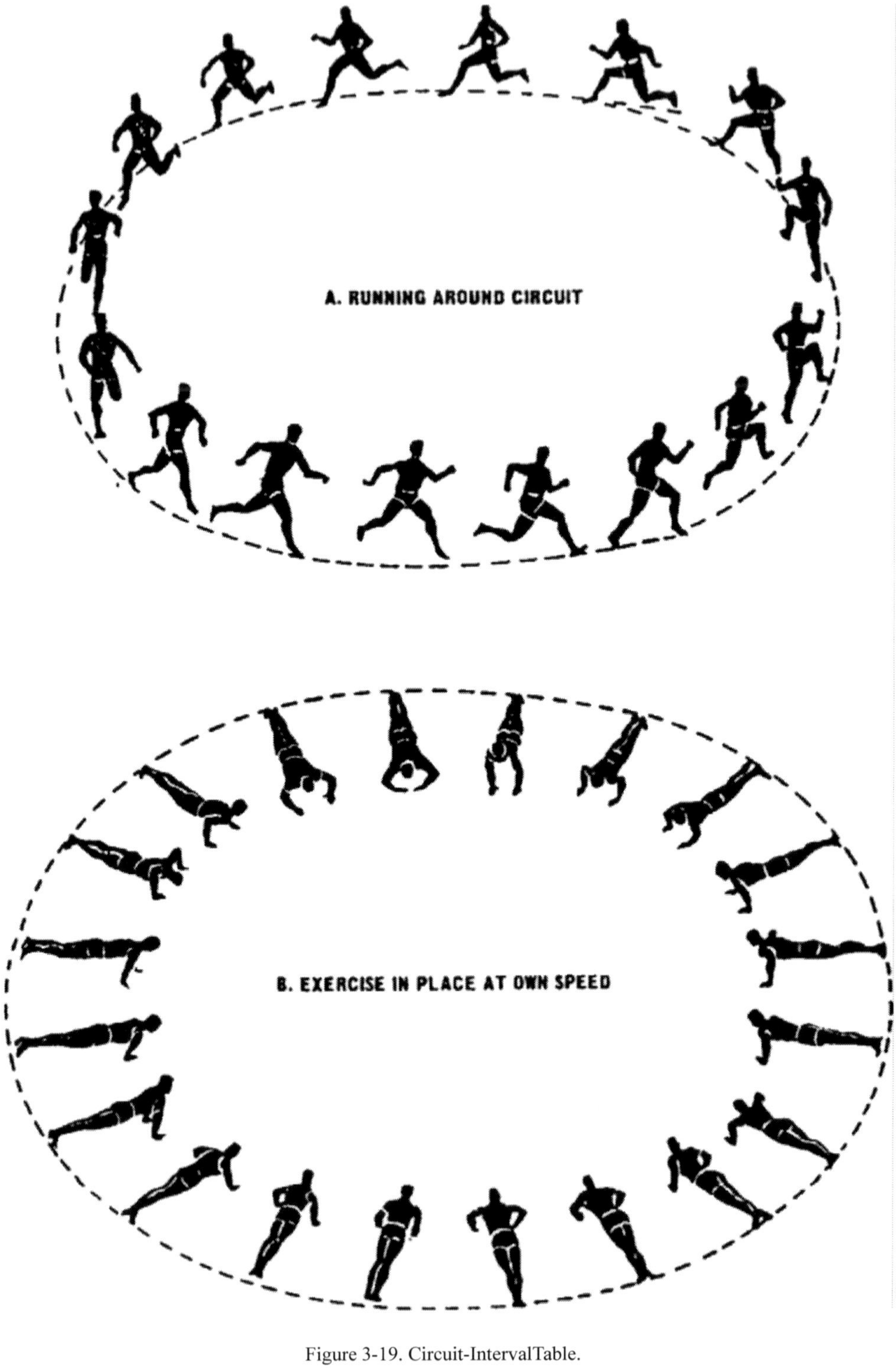

Figure 3-19. Circuit-IntervalTable.

Figure 3-20. Jumping.

Figure 3-21.Dodging.

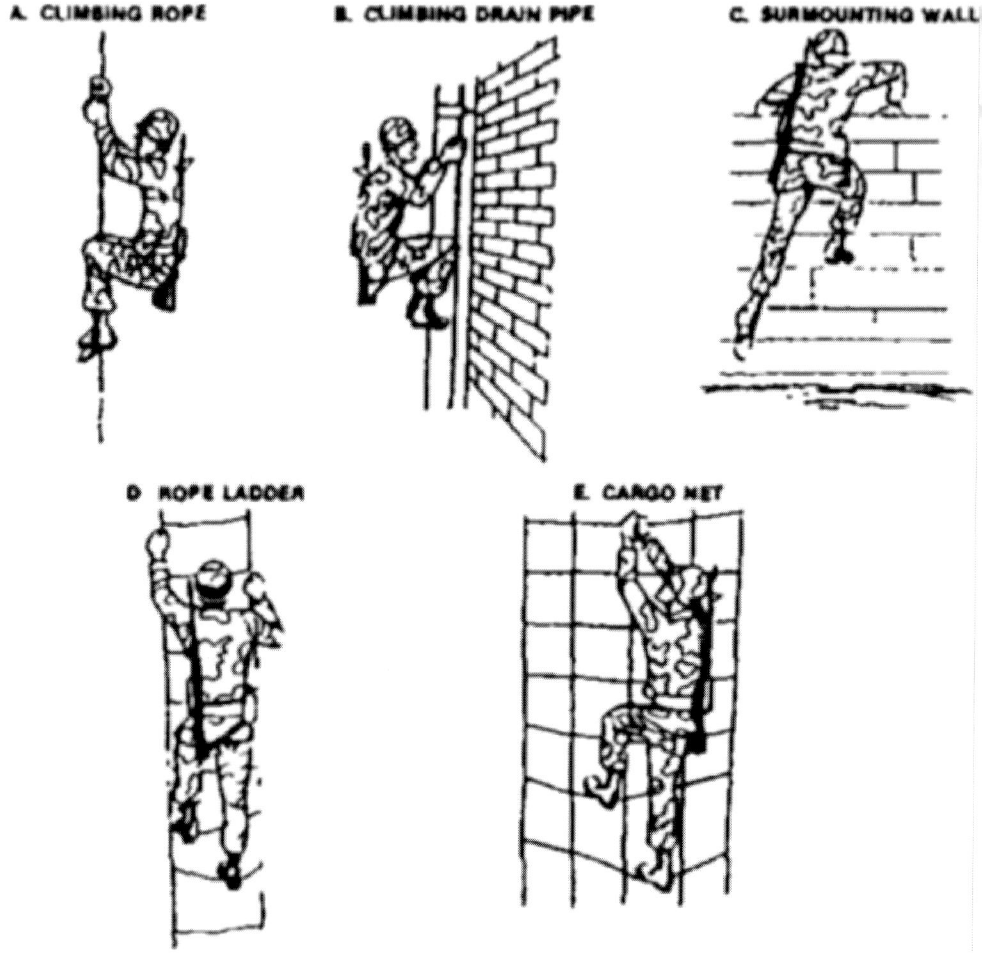

Figure 3-22.Climbing and Surmounting Vertical Objects.

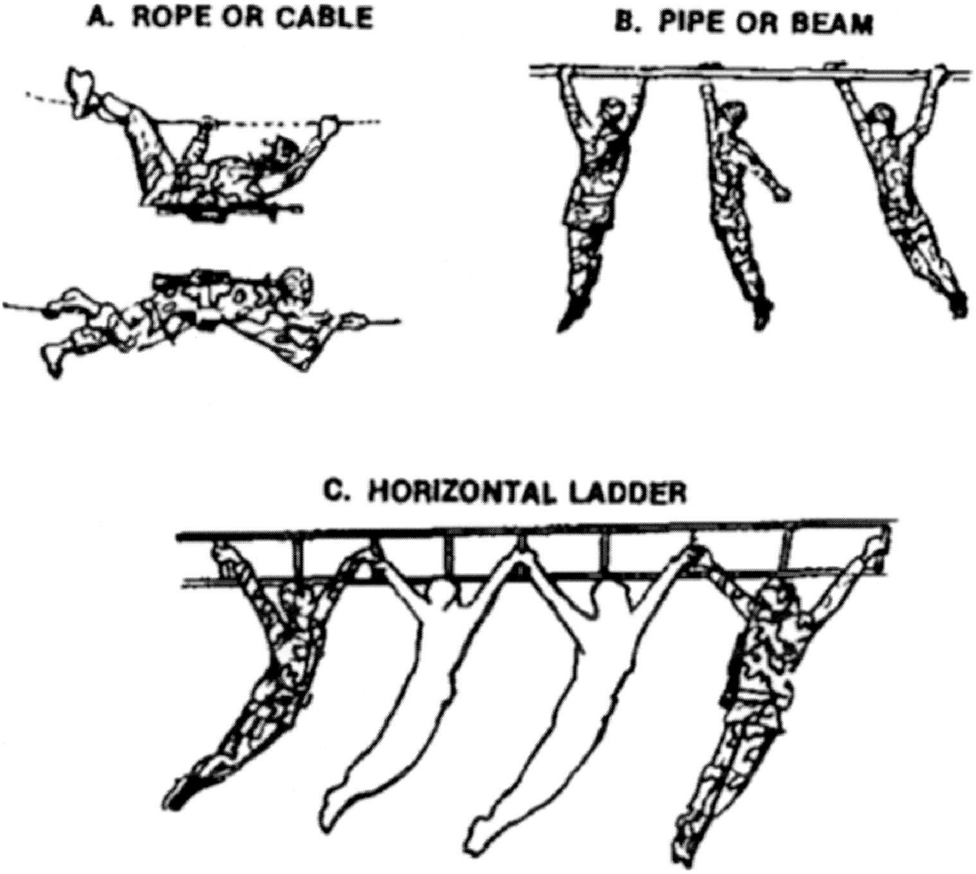

Figure 3-23. Traversing Horizontal Objects.

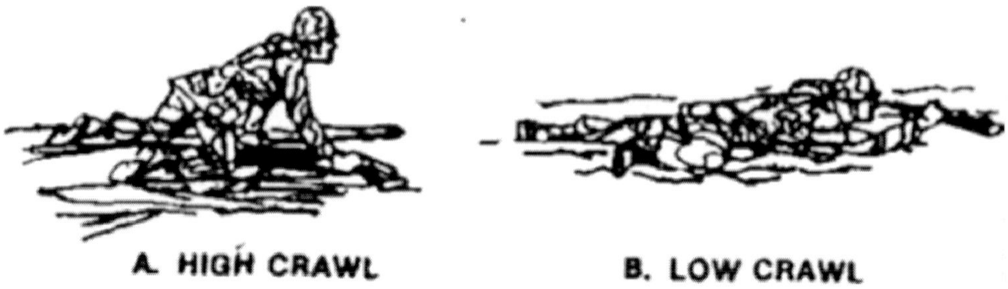

Figure 3-24.Crawling.

Figure 3-25.Throwing.

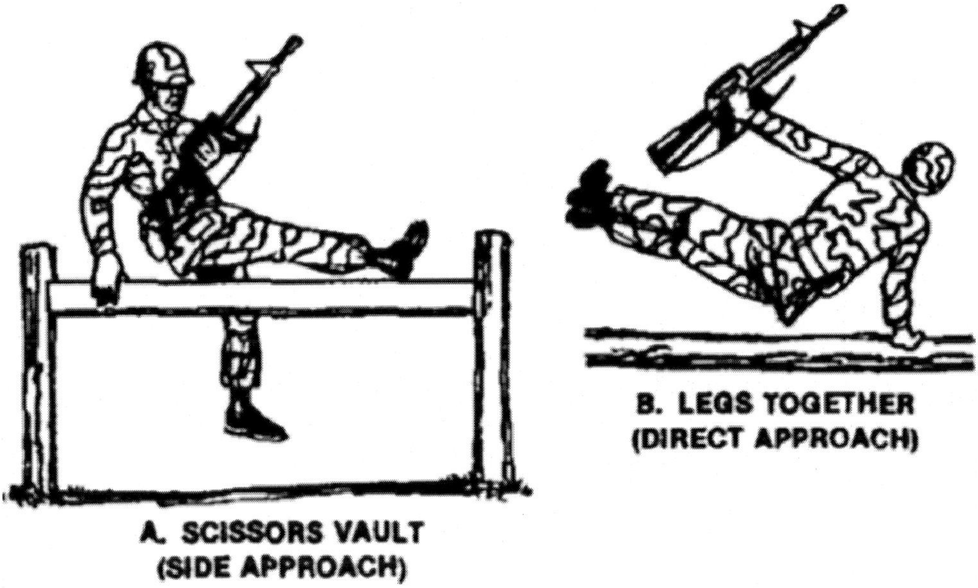

Figure 3-26. Vaulting

Figure 3-27.Balancing.

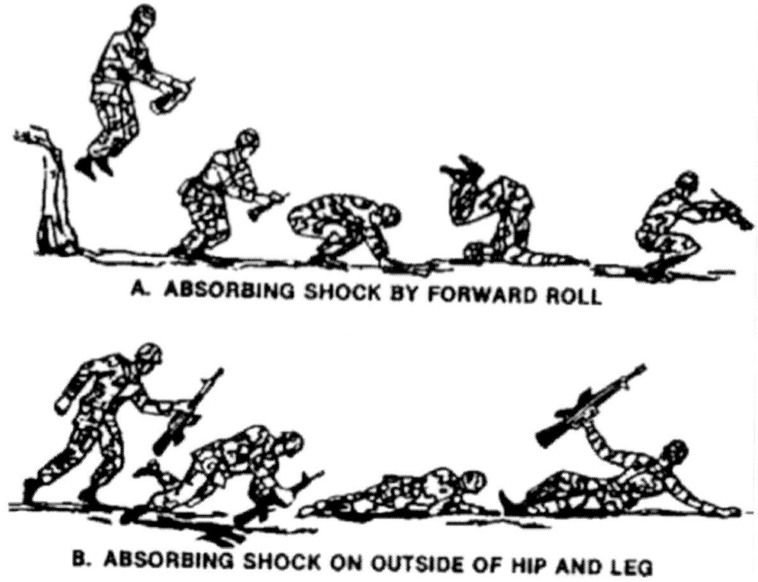

Figure 3-28.Falling.

Figure 3-29.Jumping-Type Obstacles.

LANES TO GUIDE CHANGE OF DIRECTION

MAZES TO CAUSE CHANGE OF DIRECTION

Figure 3-30.Dodging-Type Obstacles.

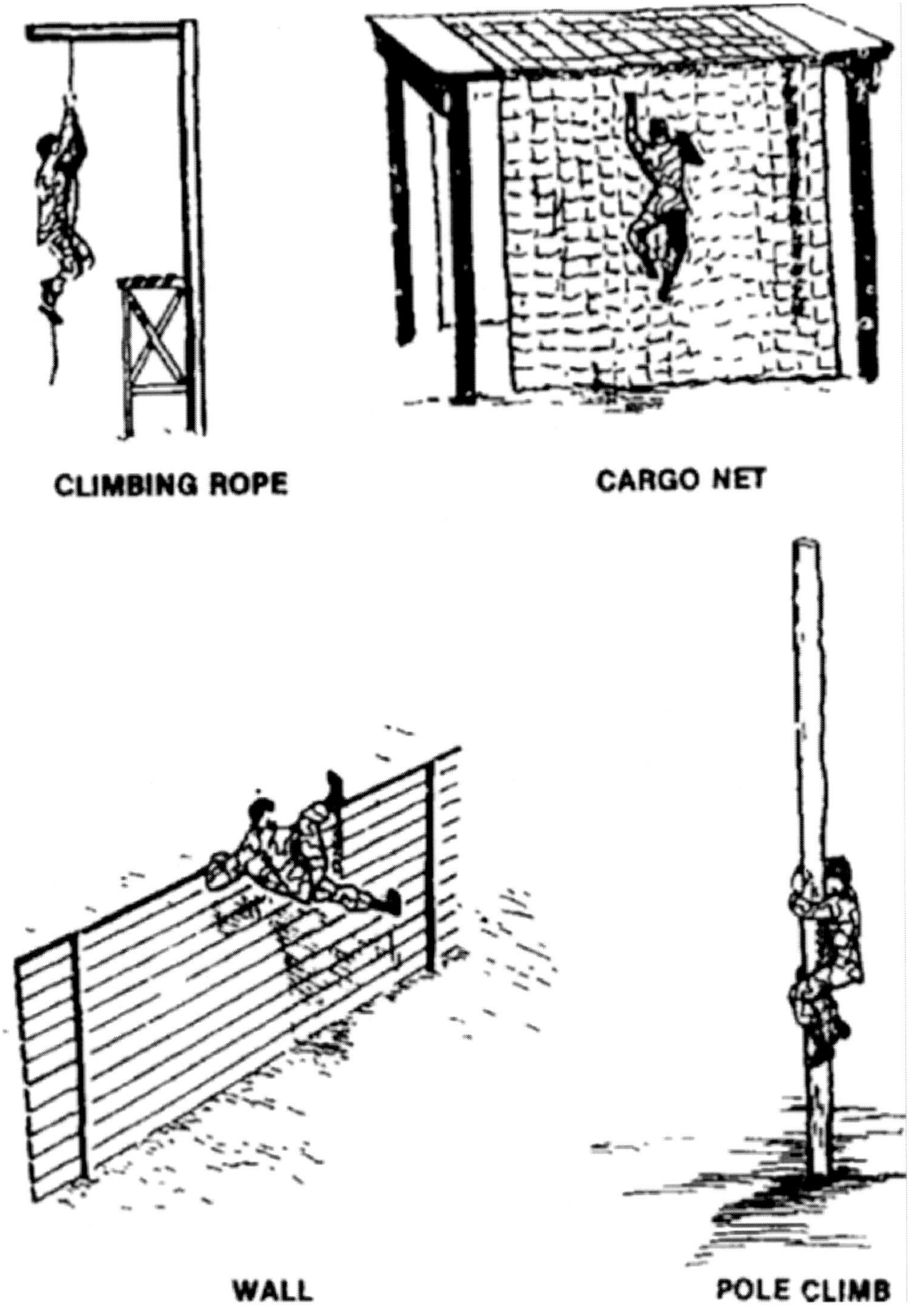

Figure 3-31. Vertical Climbing Obstacles.

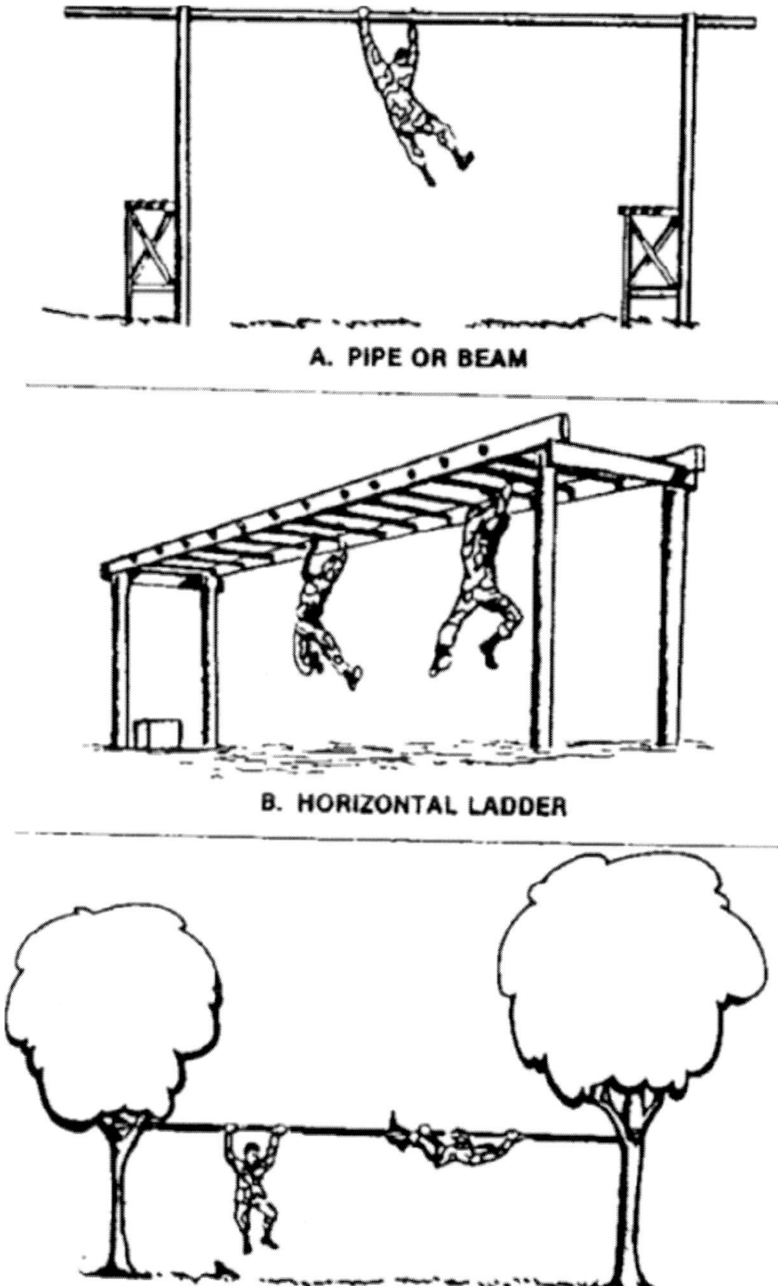

Figure 3-32.Horizontal Type Obstacles.

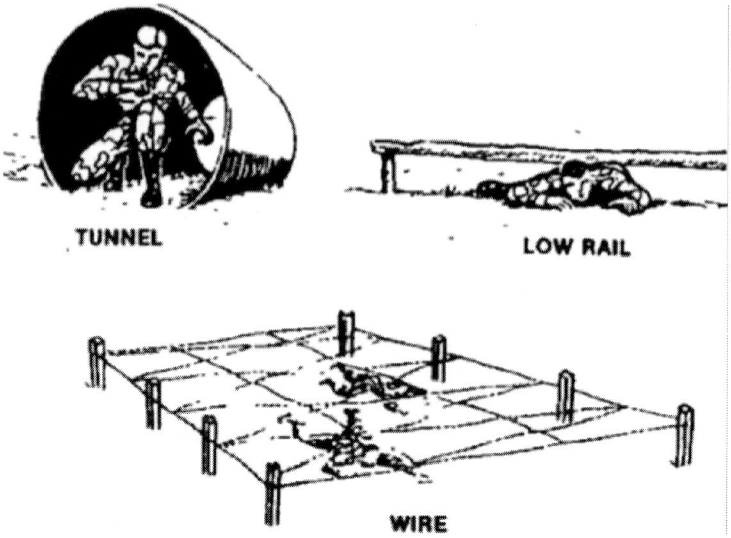

Figure 3-33.Crawling-Type Obstacles.

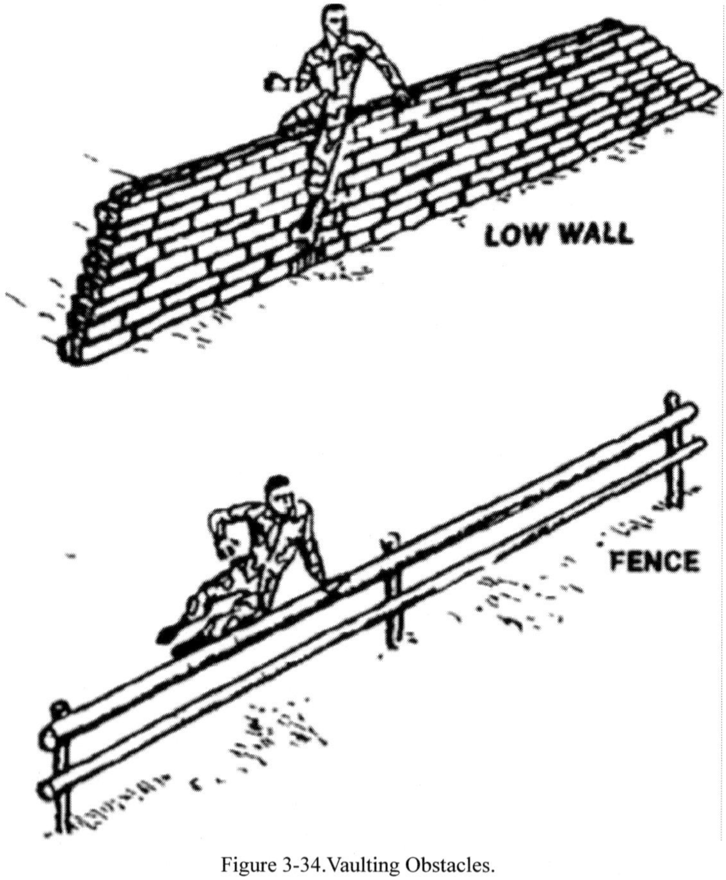

Figure 3-34.Vaulting Obstacles.

Figure 3-35.Balance-Type Obstacles.

A. THE BELLY BUSTER

B. REVERSE CLIMB

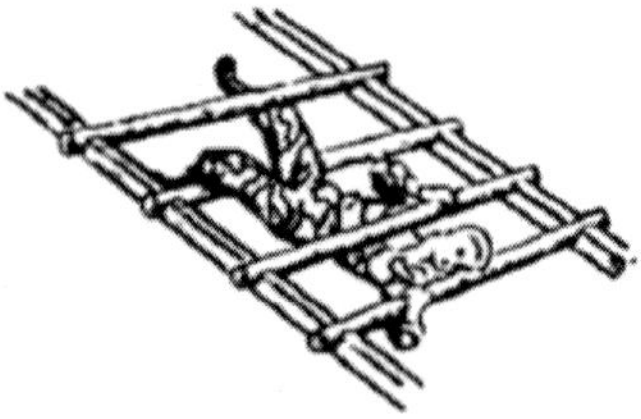

C. THE WEAVER

D. HIP—HIP

E. BALANCING LOGS

F. ISLAND HOPPER

Figure 3-36.Red Group.

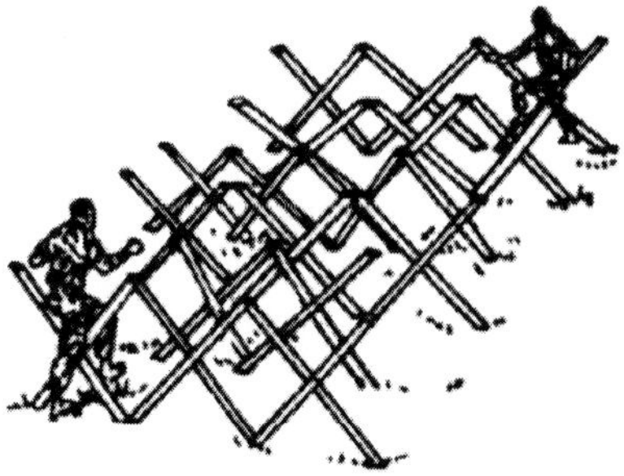

A. THE TOUGH NUT

B. SLIDE FOR LIFE

C. LOW BELLY OVER

D. BELLY CRAWL

E. THE DIRTY NAME

F. THE TARZAN

Figure 3-37.White Group.

Figure 3-38.Blue Group.

A. INCLINE WALL

B. SKYSCRAPER

C. JUMP AND LAND

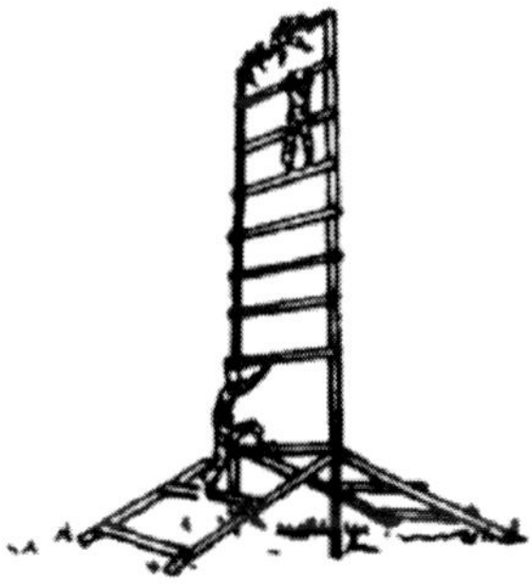

D. CONFIDENCE CLIMB

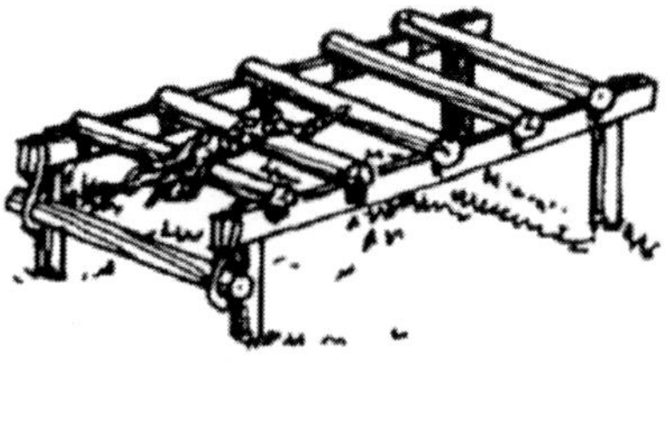

E. BELLY ROBBER

F. THE TOUGH ONE

Figure 3-39.Black Group.

Figure 3-40. Bench Conditioning, Table I.

Figure 3-41. Bench Conditioning, Table II.

PROGRESSION GUIDE

AGE GROUP	LEVEL	EXERCISES 1	2	3	4	5	6
17	A	15	18	14	15	15	250
to	B	13	16	13	13	13	235
29	C	11	14	12	11	11	[illegible]
30	A	13	14	12	13	13	200
to	B	11	13	11	11	11	185
39	C	9	12	10	9	9	165
40	A	11	11	10	11	11	150
to	B	9	10	9	9	9	135
44	C	7	9	8	7	7	120
45	A	9	8	8	9	9	100
to	B	7	7	7	7	7	90
49	C	5	6	6	5	5	80
50	A	7	6	6	7	7	75
to	B	5	5	5	5	5	70
59	C	3	4	4	3	3	60
60	A	4	5	4	4	4	50
and	B	3	4	3	3	3	40
over	C	2	3	2	2	2	30
Minutes for each exercise		2	1	1	1	2	5

Table I. Progression Guide.

Table I. Exercises.

PROGRESSION GUIDE

AGE GROUP	LEVEL	EXERCISES 1	2	3	4	5	6
17	A	17	17	17	9	19	300
to	B	15	15	15	8	17	270
29	C	13	13	13	7	15	245
30	A	15	15	15	8	17	235
to	B	13	13	13	7	15	210
39	C	11	11	11	6	13	190
40	A	13	13	13	7	15	175
to	B	11	11	11	6	13	155
44	C	9	10	9	5	11	135
45	A	11	11	11	6	13	125
to	B	9	9	9	5	11	110
49	C	7	7	7	4	9	100
50	A	9	9	9	5	11	95
to	B	7	7	7	4	9	85
59	C	5	5	5	3	7	75
60	A	6	7	7	4	9	70
and	B	5	5	5	3	7	60
over	C	4	4	4	2	5	50
Minutes for each exercise		1	1	1	1 1/2	1 1/2	6

Table II. Progression Guide.

Table II. Exercises.

PROGRESSION GUIDE

AGE GROUP	LEVEL	EXERCISES 1	2	3	4	5	6
17	A	10	19	19	16	10	350
to	B	9	17	17	15	9	315
29	C	8	15	15	14	8	280
30	A	9	17	17	14	9	270
to	B	8	15	15	13	8	240
39	C	7	13	13	12	7	210
40	A	8	15	15	12	8	200
to	B	7	13	13	11	7	180
44	C	6	11	11	10	6	160
45	A	7	13	13	10	7	150
to	B	6	11	11	9	6	135
49	C	5	9	9	8	5	120
50	A	6	11	11	8	6	115
to	B	5	9	9	7	5	105
59	C	4	7	7	6	4	95
60	A	5	9	9	7	5	90
and	B	4	7	7	6	4	80
over	C	3	5	5	4	3	70
Minutes for each exercise		1 1/2	1	1	1 1/2	1	6

Table III. Progression Guide.

Table III. Exercises.

PROGRESSION GUIDE

AGE GROUP	LEVEL	EXERCISES 1	2	3	4	5	6
17	A	12	9	12	24	25	400
to	B	11	8	11	22	23	380
29	C	10	7	10	21	21	360
30	A	11	8	11	23	23	305
to	B	10	7	10	21	21	290
39	C	9	6	9	20	20	275
40	A	10	7	10	20	21	225
to	B	9	6	9	18	18	215
44	C	8	5	8	16	16	205
45	A	8	6	8	16	16	175
to	B	7	5	7	14	14	165
49	C	6	4	6	12	12	155
50	A	6	5	6	13	13	135
to	B	5	4	5	11	11	130
59	C	4	3	4	10	10	120
60	A	5	4	5	10	10	100
and	B	4	3	4	9	9	95
over	C	3	2	3	8	8	90
Minutes for each exercise		1	2	1	1	1	6

Table IV. Progression Guide.

Table IV. Exercises.

PROGRESSION GUIDE

AGE GROUP	LEVEL	EXERCISES 1	2	3	4	5	6
17	A	14	13	28	14	30	450
to	B	13	12	27	13	28	430
29	C	12	11	26	12	26	410
30	A	12	12	25	12	26	350
to	B	11	11	24	11	24	330
39	C	10	10	23	10	22	310
40	A	11	11	23	11	23	250
to	B	10	10	21	10	21	240
44	C	9	9	19	9	19	230
45	A	9	9	20	9	20	200
to	B	8	8	18	8	18	190
49	C	7	7	16	7	16	180
50	A	7	7	16	7	16	170
to	B	6	6	14	6	14	155
59	C	5	5	12	5	12	140
60	A	6	6	12	6	12	115
and	B	5	5	11	5	10	110
over	C	4	4	9	4	9	105
Minutes for each exercise		2	1	1	2	1	5

Table V. Progression Guide.

Table V. Exercises.

PROGRESSION GUIDE

AGE GROUP	LEVEL	EXERCISES 1	2	3	4	5	6
17	A	17	15	32	32	35	500
to	B	16	14	30	30	33	480
29	C	15	13	28	28	31	460
30	A	15	13	30	30	31	400
to	B	14	12	28	28	29	380
39	C	13	11	26	26	27	360
40	A	13	10	27	27	27	310
to	B	12	9	25	25	25	285
44	C	11	8	23	23	23	265
45	A	11	9	23	23	23	250
to	B	10	8	21	21	21	230
49	C	9	7	19	19	19	210
50	A	9	8	19	19	19	200
to	B	8	7	17	17	17	190
59	C	7	6	15	15	15	175
60	A	8	7	15	15	17	140
and	B	7	6	13	13	15	130
over	C	5	5	10	10	12	120
Minutes for each exercise		2	1	1	1	1	6

Table VI. Progression Guide.

Table VI. Exercises.

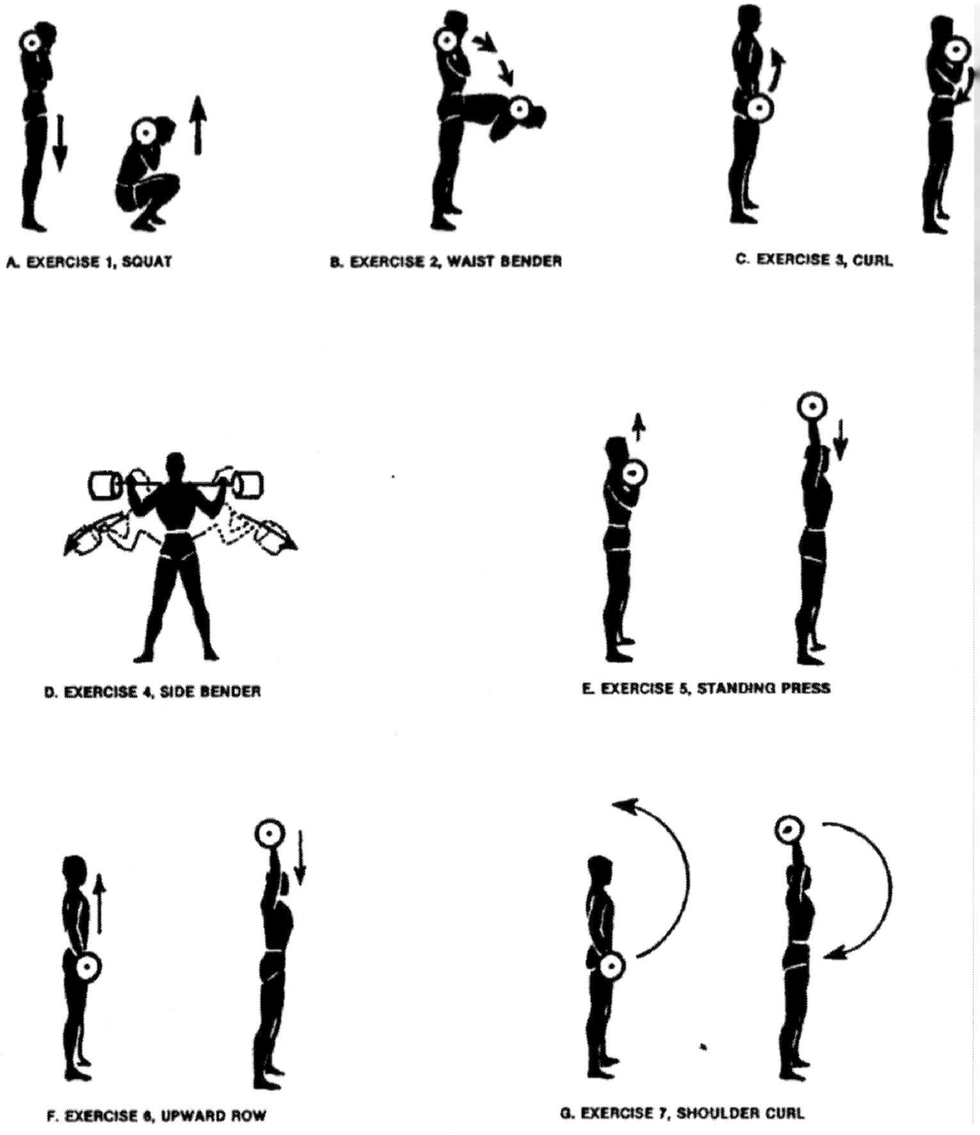

Figure 3-42. Weight Training.

Figure 3-43. Isometric Exercises.

Figure 5-1.Combatives Table I.

Figure 5-2.Combatives Table II.

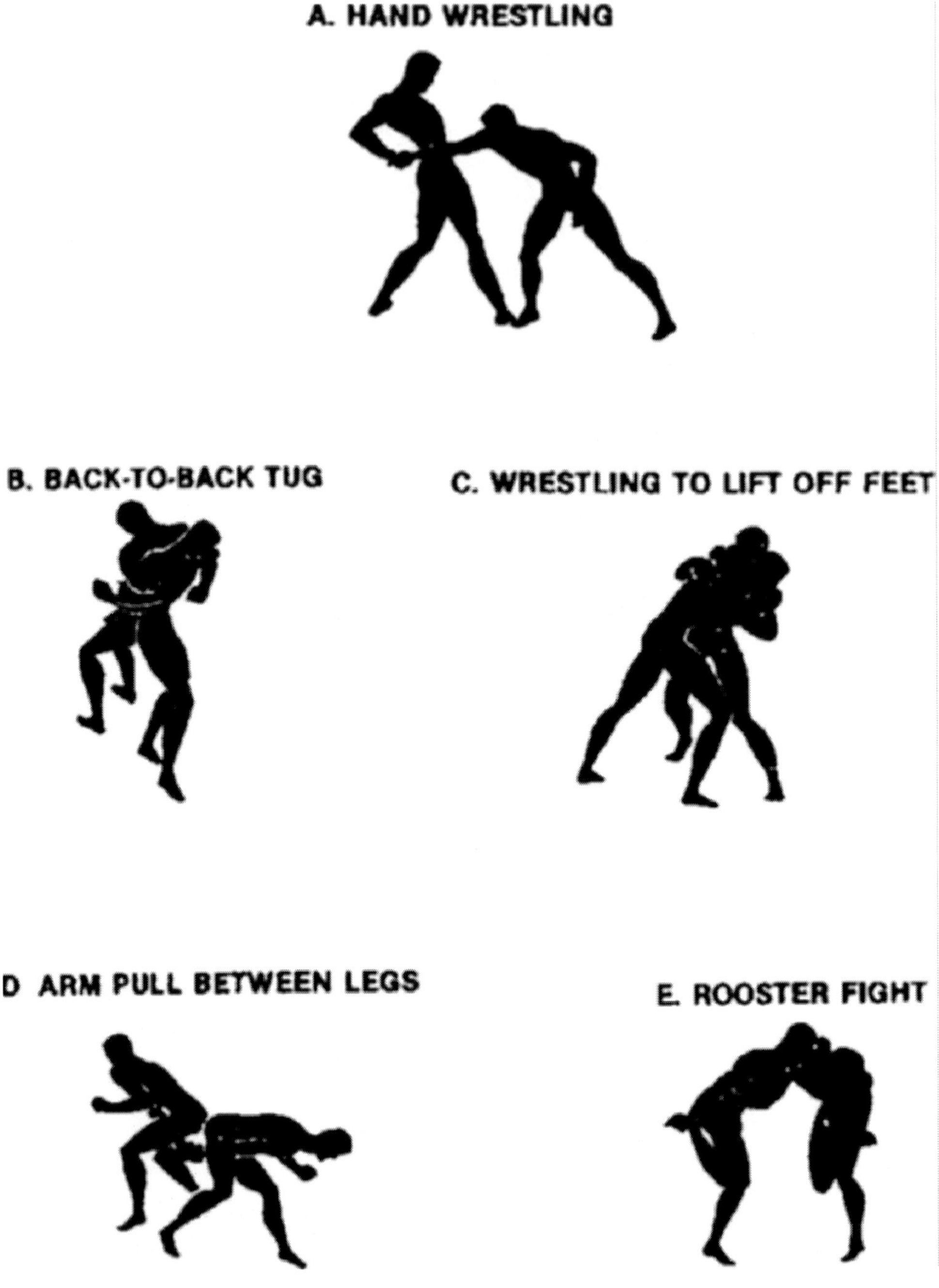

Figure. 5-3.Combatives Table III.

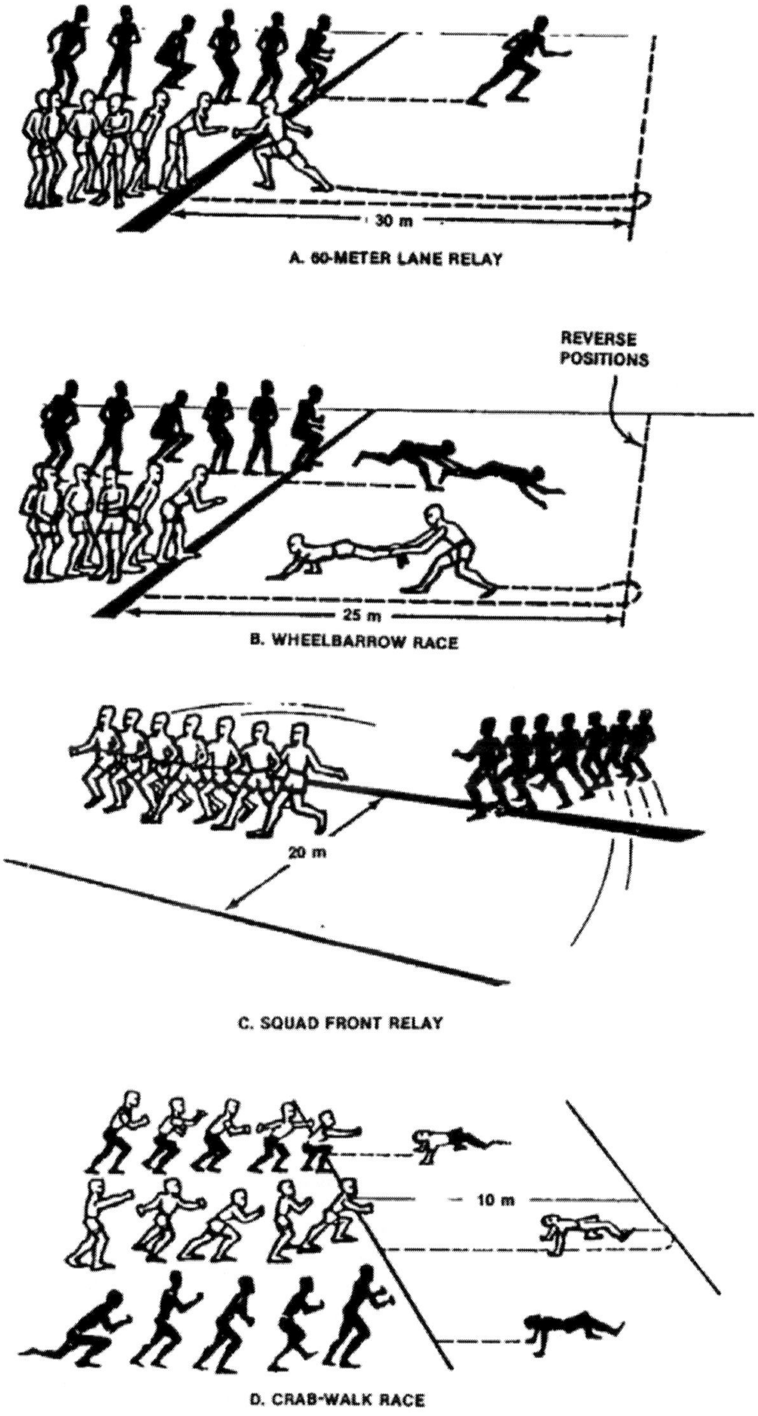

Figure 5-4.Relay Table I.

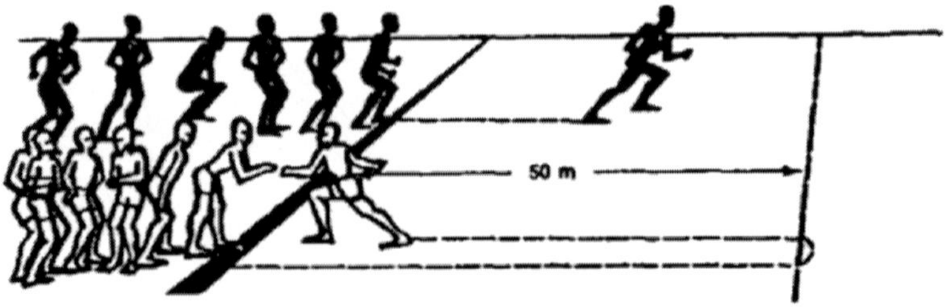

A. 100-METER LANE RELAY

B. FROG-JUMP RELAY

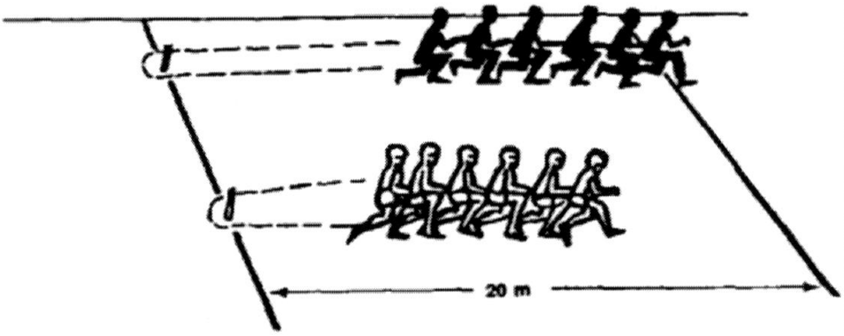

C. SIMPLE RELAY

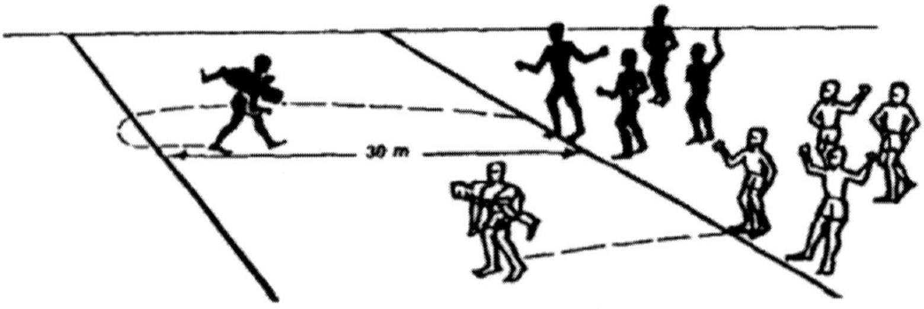

D. FIREMAN'S CARRY RELAY

Figure 5-5.Relay Table II.

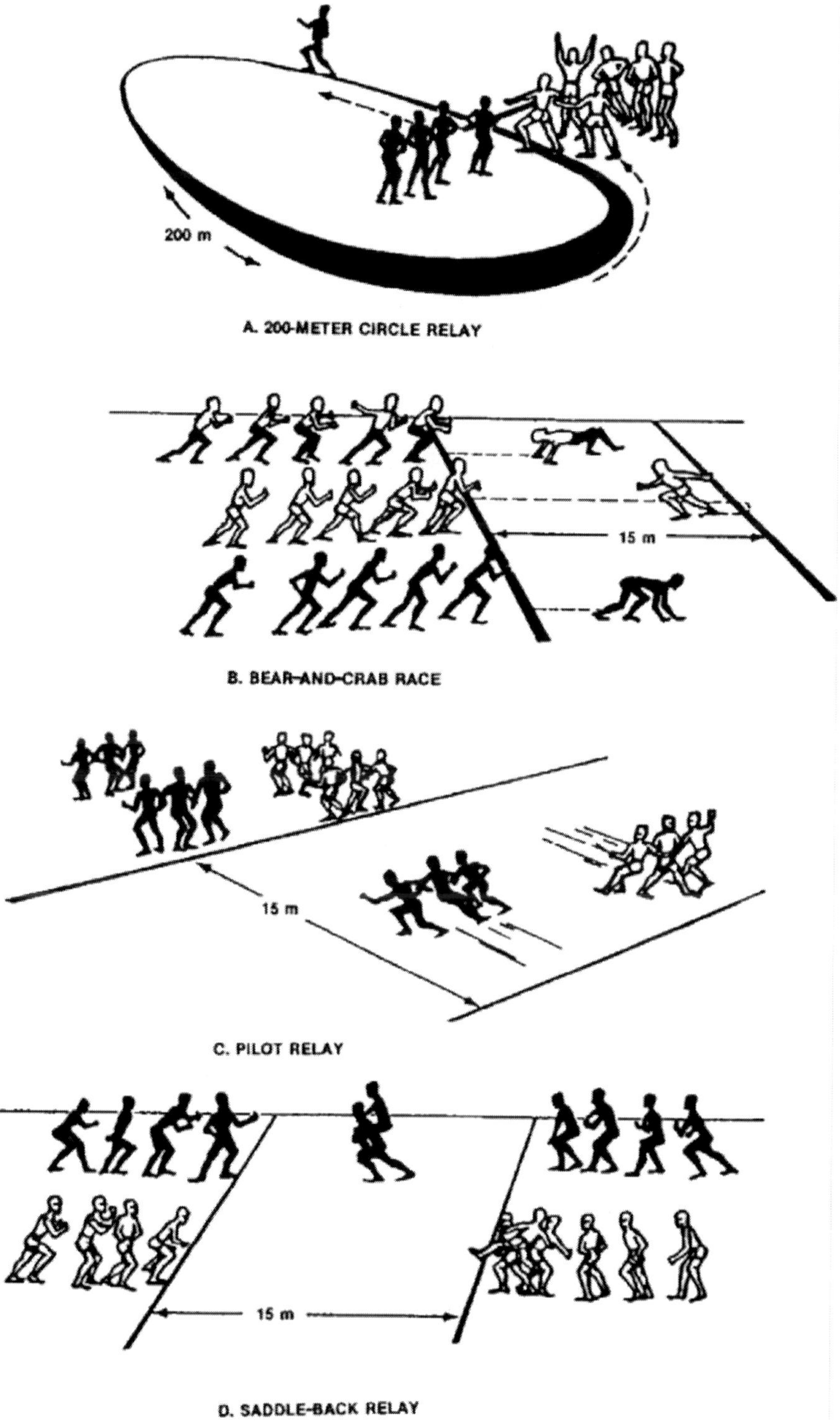

Figure 5-6.Relay Table III.

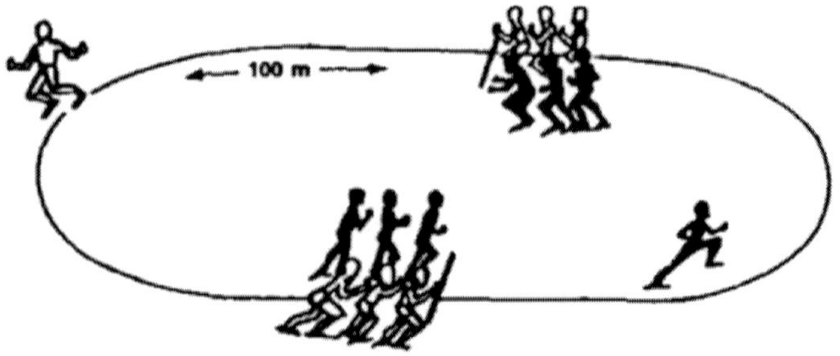

A. 100-METER CIRCLE RELAY

B. IN-AND-OUT RELAY

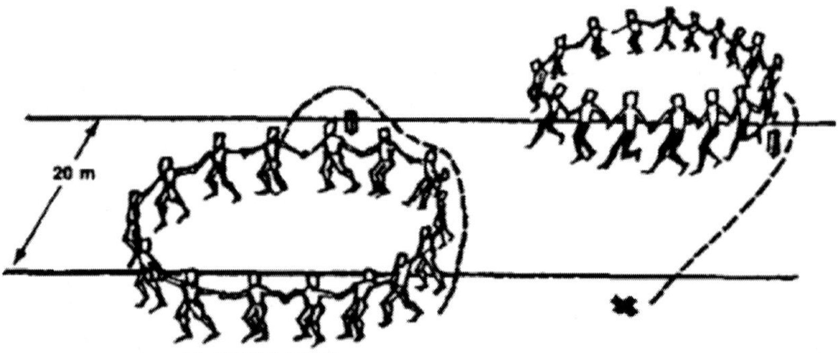

C. CIRCLE RACE

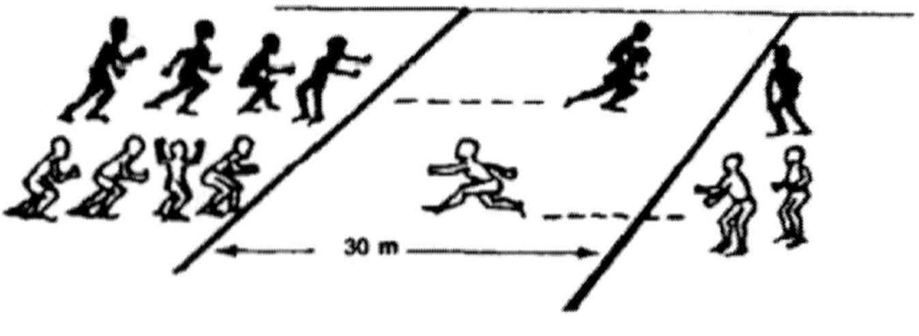

D. HORSE-AND-RIDER RELAY

Figure 5-7.Relay Table IV.

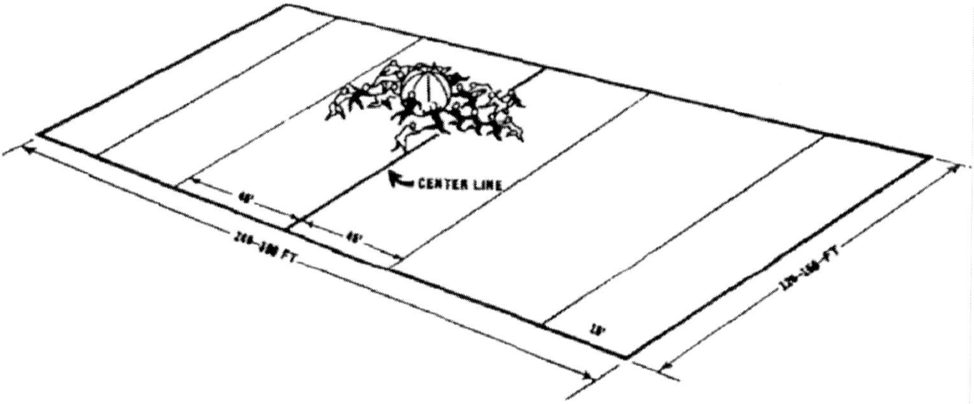

Figure 5-8.Pushball.

Figure 5-9.Line Rush.

Figure. 5-10. Human Tug-of-War.

Figure 5-11. Master of the Ring.

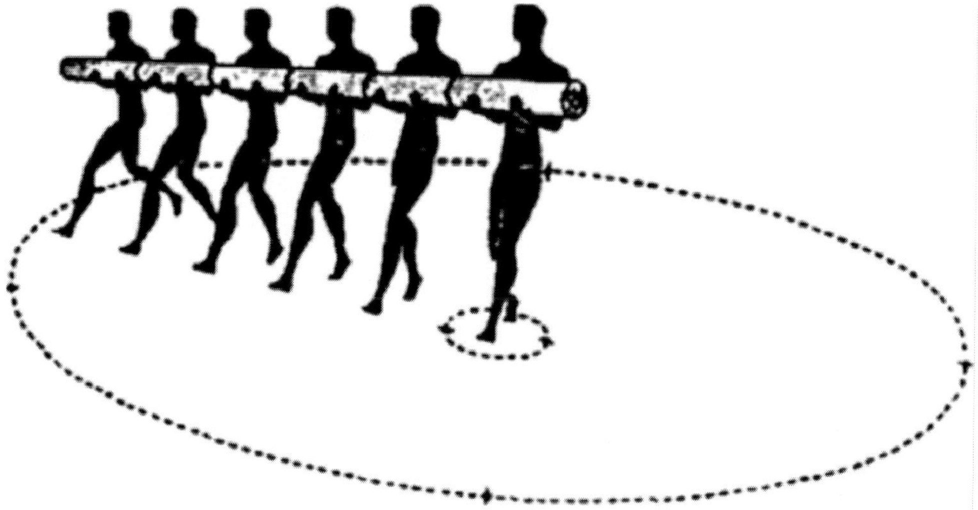

Figure 5-12.Log Pivot Circle.

Figure 5-13.Log-Rolling Race.

Figure 5-14. Prone Pushing Contest.

Figure 5-15. Shuttle Relay Race.

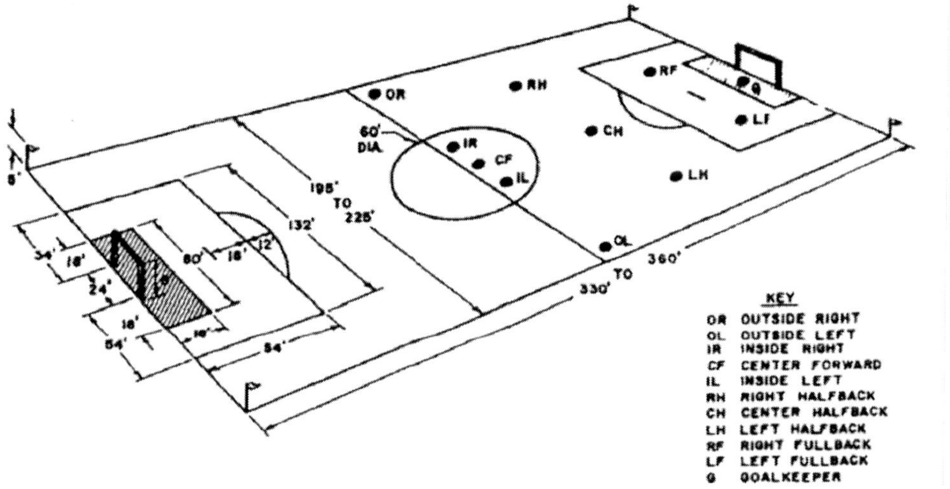

Figure 5-16. Soccer.

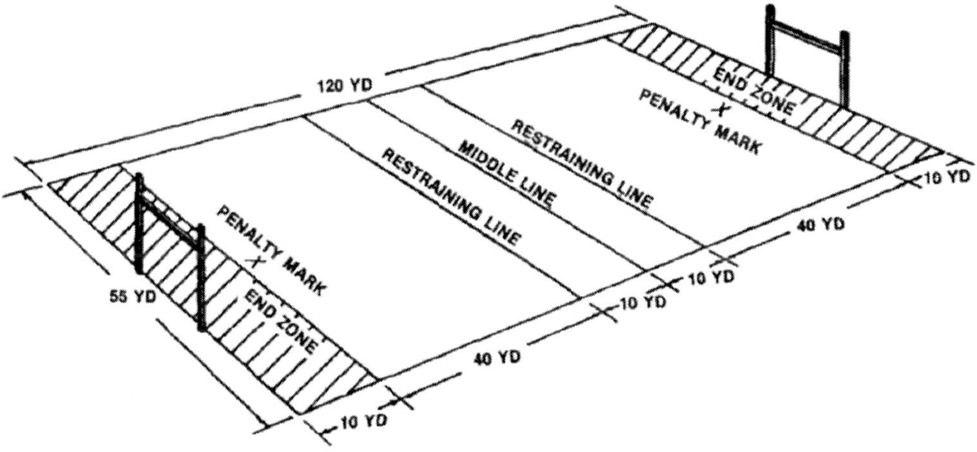

Figure 5-17.Speedball Field.

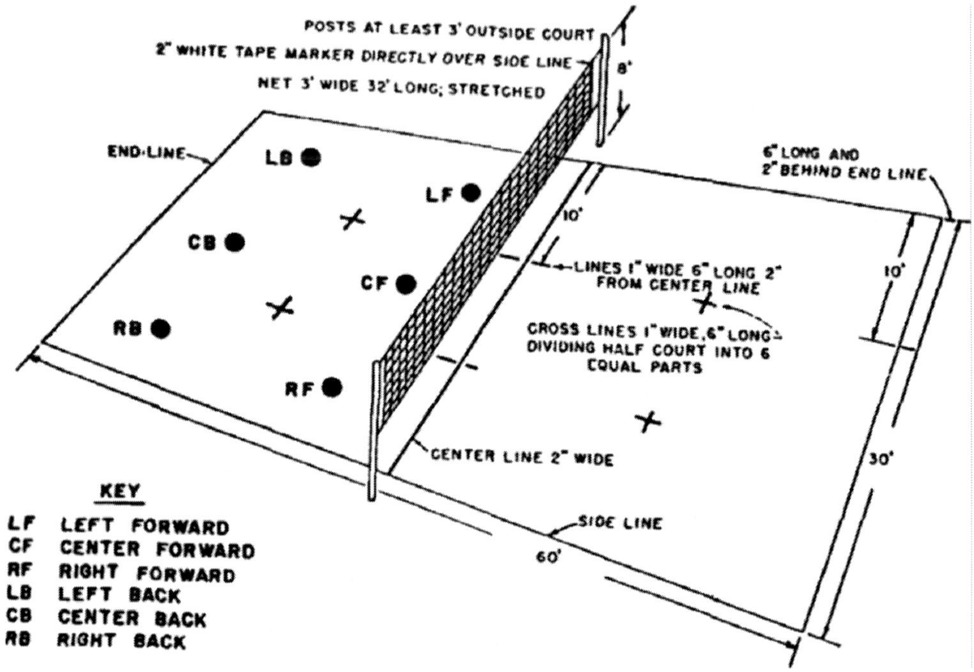

Figure 5-18.Volleyball.

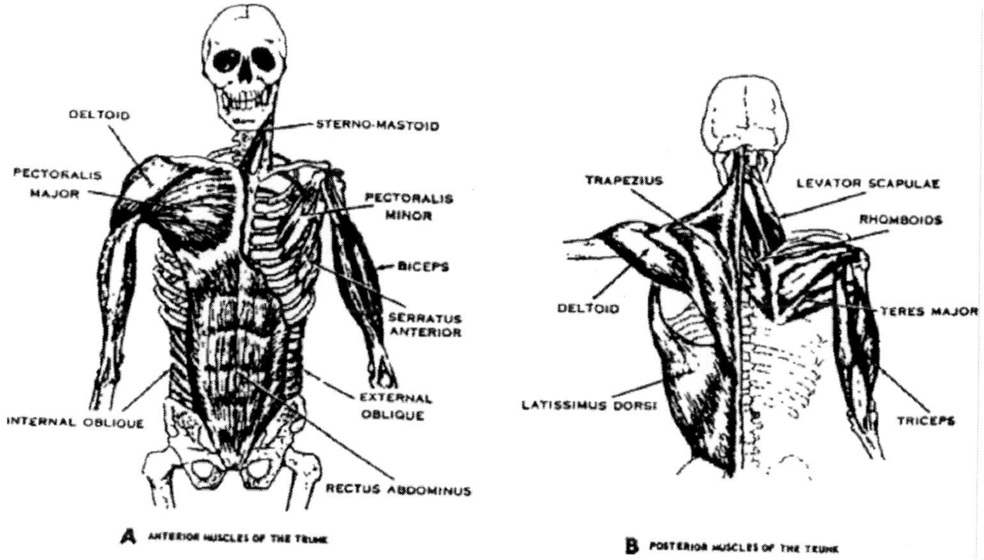

Figure 7-2.Muscles of the Trunk.

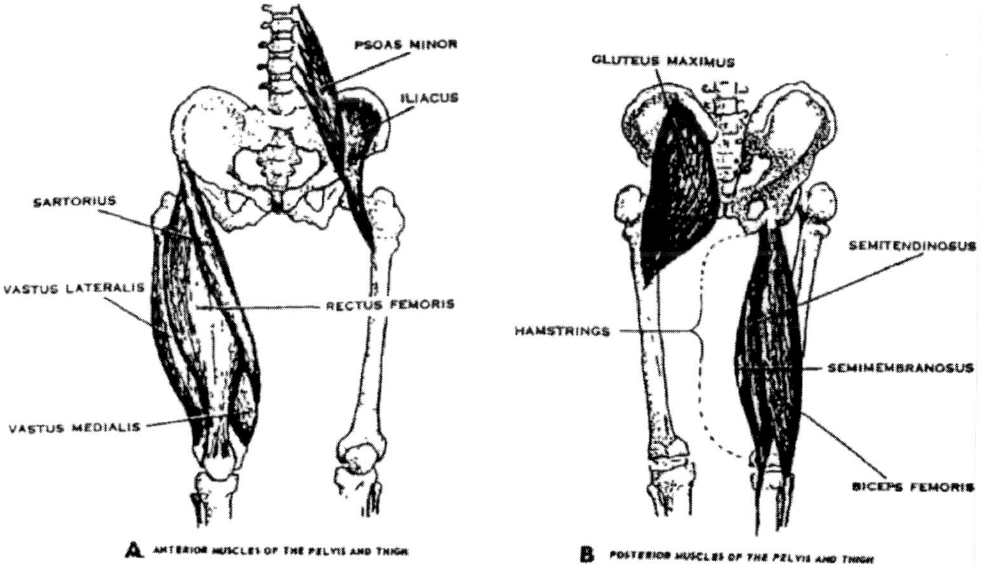

Figure 7-3.Anterior and Posterior Muscles of the Pelvisand Thigh.

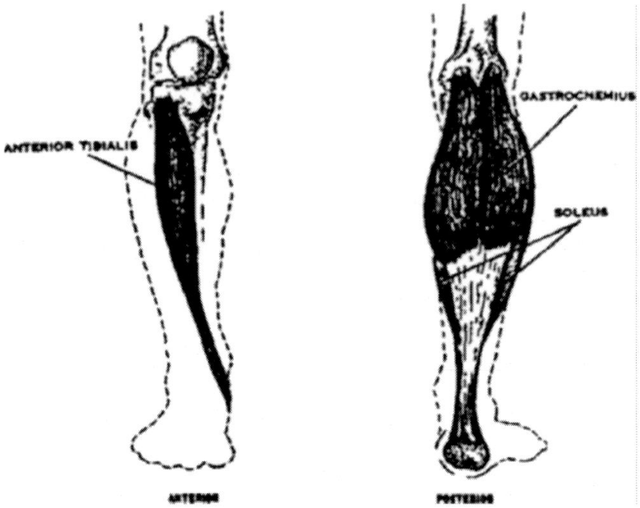

Figure 7-4. Muscles of Lower Leg.

Figure 7-5.Circulation of Blood from Heart to the Body.

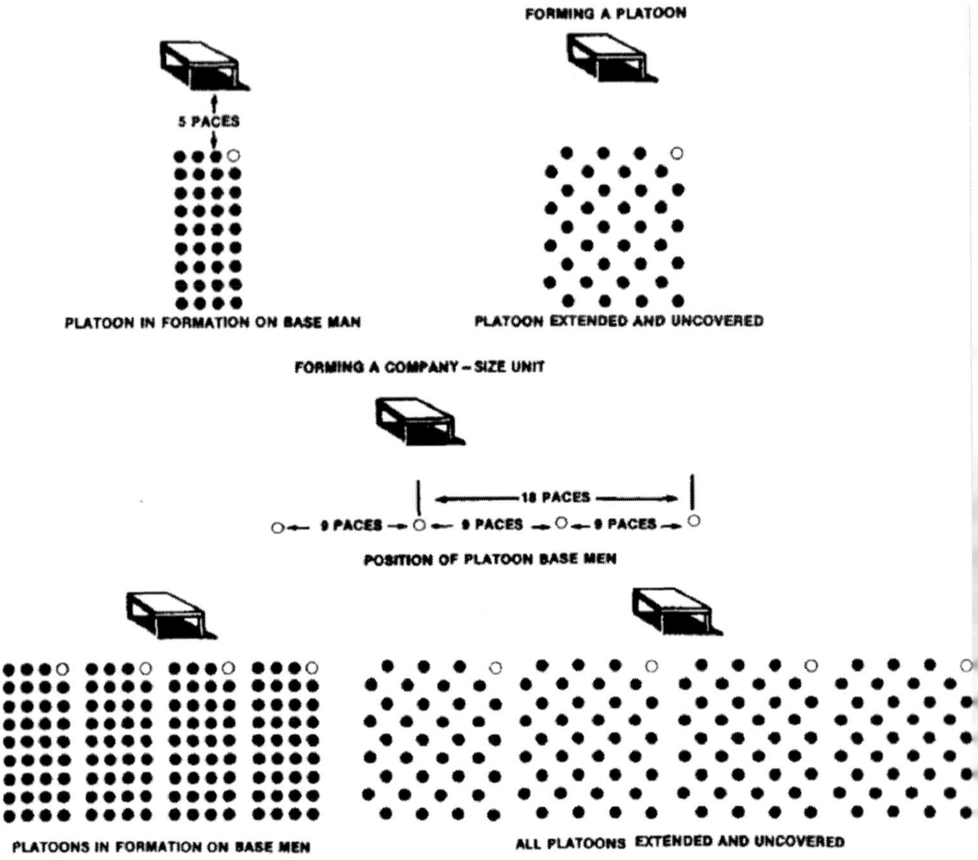

Figure A-l. Forming the Extended Rectangular Formation.

Figure A-2.Circle Formation.

U.S. Marine Corps

IRAQ An Introduction to the Country and People

Department of Defense

Pentagon
Publishing

Iraq:

An Introduction to the Country and People

Introduction

Overview

Purpose

The United States is committed to on-going involvement with Iraq to rebuild the country. Marines, who deploy to Iraq or are involved with Iraqis, need a basic knowledge of the country, its culture, history and present-day state of affairs.

[] This handbook explains a number of basic issues that should be in the knowledge "toolbox" of a Marine working with Iraqis or deployed to the region.

[] You can use the links that form an integral part of this document to explore further any of the topics discussed in this handbook.

Contents

This handbook includes the following topics:

Continued on next page

Contents, (continued)

Three Critical Factors

There are three important factors that have shaped Iraqi culture and society:

[] **Islam** - particularly, the division between Sunnis and Shias (Shi'ites)
[] **Oil** - which supplies Iraq with its "life blood"
[] **Saddam Hussein** - former leader of the country

You will learn more about each of these as you read this handbook.

Continued on next page

Overview, Continued

Geography

Iraq or Irak, officially the Republic of Iraq, a country of 167,924 square miles (434,924 sq km) in southwest Asia is bordered on the

[] South by Kuwait, the Persian Gulf, and Saudi Arabia
[] West by Jordan and Syria
[] North by Turkey
[] East by Iran

Baghdad is the capital and largest city. Other large cities are Mosul and Kirkuk and the port of Basra. The country is divided into 18 provinces.

http://www.lib.utexas.edu/maps/iraq.html has a number of good maps showing Iraq.

Climate and Weather

Most of Iraq gets little rainfall. The winters are cool; summers are dry and hot, with blowing winds and sandstorms. The exception is the northern mountains along the Iranian and Turkish borders. Here the winters are cold with heavy snows.

Flooding from the rivers because of melting snow is a problem farther south, but is less common today because of flood-control projects. Agriculture depends mainly on water from the rivers.

[] Because of the sandy soil and intense heat, Iraqi farmers in the southeast grow "orchards" of dates and cotton.

[] Farther up the rivers, as the elevation increases, rainfall allows different kinds of crops, such as grains and vegetables.

[] In the mountains of the north, agriculture gives way to oil. The largest oil fields are near Mosul and Kirkuk.

Note: Appendix B provides you with quick facts about Iraq.

(This page intentionally left blank.)

Iraqi Culture, History, and Religion

Culture and Peoples

Cultural and Ethnic Factors

Iraq with its population of about 22 million people has been shaped by a number of cultural and ethnic factors.

- []Most of the people are of Arabic origin. About 80% speak Arabic.
- [] Ninety-five percent of the people arc Muslim. There arc about twice as many Shias (Shi'ites) as Sunnis. (Sunnis are the more numerous sect or religious group in most Muslim countries.)
- []Kurds are a large minority living in the uplands of northeast Iraq. They are primarily Sunni Muslims.

Minority Groups

Other minorities of Iraq include Turks, Armenians, and Assyrians (Nestorian Christians). Most of the country's once large Jewish population resettled in Israel in the early 1950s.

Urban and Rural

Iraq's only way of reaching the sea is a short stretch of coast on the northwestern end of the Persian Gulf, including the Shatt al Arab waterway.

For the majority of the population, life centers on the Tigris and the Euphrates rivers, which flow southward and come together in the Shatt al Arab. Urban growth is in this area. There are also numerous marshlands and water basins between these two rivers. The southwest is desert and supports a small population of nomad shepherds. Present day Iraq encompasses the approximate area of the Middle East called Mesopotamia in ancient times.

Continued on next page

Selected Topics The topics covered in this section on Iraqi history, culture, and peoples arc shown in the table below.

Kurds

Special Case

The Kurds are one among a number of minorities living in present-day Iraq. Because of their oppression by the regime of Saddam Hussein, which is covered later in this handbook, you will read about these people separately in this section.

Ancient History: Key to the Present

The Kurds are an old, Indo-Iranian tribe who lived in ancient Persian times in areas that the present-day Kurds, descendents of this tribe, still occupy. This area is often called Kurdistan.

A Distinct Name

The name *Kurd* was used around 640 B.C. It described a group of nomads that lived on the plateaus and mountains between Armenia and the Zagros Mountains. Later, the Arabs applied the name to the residents of these mountainous areas who had converted to Islam.

Kurdish Expansion

[]Eventually, the Kurds spread out and settled in a region reaching from central Anatolia (modern-day Turkey) to northeast Iran.

[]The Kurds have developed as "isolated" units, physically cut off from one another into what is often called the "Kurdish diaspora."

Current Population Distribution

Currently, most of the Kurdish people, numbering about 25 million total, live in Turkey, Iraq, Syria, and Iran, primarily.

[]About half of this total population lives in Turkey.
[]6-7 million live in Iran.
[]3.5-4 million, in Iraq.
[] 1.5 million live in Syria.

There are groups living outside the Middle East in Armenia, Germany, Sweden, France, and the United States. Kurdish communities also exist in various countries of the former Soviet Union.

The Kurds in Iraq

The Iraqi Kurds make up about 23% of Iraq's population. They live near the towns of Mosul, Kirkuk, and Sulaimaniyah. The Iraqi Kurds are sometimes known by the name Failili Kurds.

Continued on next page

People vice Nation

In spite of the fact that the Kurds as an ethnic group are spread over such a large area, most of them are bound together as a common group by their language, traditions, and their culture. In recent interviews among Kurds living in southeastern Turkey, respondents considered themselves to be Kurds first, Turks second.

The concept of a Kurdish sovereign state is still not a full reality in the minds of all Kurds. It is certainly not an idea that the nation states of the present-day Middle East -Turkey, Iran, and Iraq - want to consider. Their hope is to "integrate" the Kurds into the respective country that they live in.

(See http://ethnicitv.vour-directorv.com/Finding the Kurds a Way Kurdistan and the Discourse of the Nation State'2242643.html for a detailed discussion of the implications of a Kurdish sovereign state.)

Indo-European, not Arabic

The Kurds are not an Arabic speaking people. They don't speak a language that is related to either Arabic or Turkish. The Kurdish language can probably be traced back to an ancient language that was more closely related to Farsi (the current language spoken in Iran) and Hindi (the current official language of India). Present-day Kurdish is closely related to the northwestern group of Iranian languages.

Religion

The Kurds accepted Islam from the Arabs during the 7 century A.D. They are Sunni Muslims, not Shias (Shi'ites).

Culture

Kurds have their own music, poetry, and dance. Individual villages have their own dances in which men and women often dance together. Their legends are filled with heroism and romance. Nature, particularly the mountains, has shaped Kurdish life.

[] Mountain flowers have contributed to the color and patterning in Kurdish clothing and kilims (woven rugs).

[] Men still wear turbans and baggy pants (called shalvari).

[] Women wear headscarves, long colorful dresses, and coats decorated with silver and gold thread.

Continued on next page

Kurds, Continued

Economy and Livelihood

The Kurds' economy has been based on agriculture from ancient times.

- About a third of their land is fit for farming.
- Many Kurds grow wheat and other grains.

In the higher mountains, the Kurds herd sheep. In the 8th century, Kurdish dairy products were considered so delicious that the Vikings from Scandinavia and medieval Russia traveled south to the Iraqi region to buy butter from Kurdish peasants.

Because of unstable conditions in present-day Iraq agriculture has declined.

A Distinct Difference

There is a distinct difference between the Arab-speaking majority of the "desert lowlands" of Iraq and the Kurdish mountain folk.

- Language separates them.
- Culture has developed along separate lines.
- Economically, the Kurds have held to their farming traditions; the Iraqi Arabs have expanded into marketing entrepreneurship.

See hUp://ethnicitv.vour-directorv.coiTi/Meet the Kurds'2242666.html for a good detailed introduction to the culture of the Kurds.

hup://www.bail]ebv.conv6.S/ku/Kurds.him] gives a good overview of Kurds.

http://www.kelim-art.de/exhib/kt/e kt index.html is a site that shows the beauty and craftsmanship of Kurdish kilims [rugs].

Iraqi Economy and Livelihood

Oil

The chief factor thai dominates Iraq's economy is oil. Traditionally it provided about 95% of the money received from foreign trade. Oil exports have been substantially reduced because of

[] The United Nations imposed economic sanctions after Iraq's invasion of Kuwait.

[] The destruction of oil fields by Saddam Hussein when the U.S. coalition invaded in 2001.

[] Current terrorist sabotage of the oil pipelines.

Oil Production

The Iraq Petroleum Company, a national company, used to produce the oil. The oil is pumped to Turkey, Lebanon, Syria, and the Persian Gulf.

[] In the early 1980s, oil production averaged about 2.0 million barrels per day (bpd).

[] In the late eighties about 1.7 million barrels per day (bpd) were pumped.

Natural Gas

Iraq also has reserves of natural gas. In 1987, it produced about 7 million cubic meters. Natural gas reserves are nearly 850 billion cubic meters.

Continued on next page

Iraqi Economy and Livelihood, continued

Other Goods and Services

Other industries produce chemicals, textiles, cement, food products, construction materials, leather goods, and machinery. Production in electronics, fertilizers, and sugar had recently started. Since the invasion of Iraq the production of these goods and services is at a standstill.

[]Services contributed about 40% to the gross domestic product.
[]Mining and manufacturing, about 6%.
[] Construction employed about 20% of the civilian population and the military forces.

The University of Texas at Austin has good detailed information about Iraq at http://inic.utexas.edu/menic/Countries and Regions/Iraq/.

http://wwvv.iraqresearch.com/ is a good Congressional Research Service site that has links about Iraq.

Labor Force

Government figures showed the industrial labor force was at 170,000 in 1984, with

[]80 percent of workers in state factories
[] 13 percent in the private sector
[]7 percent in the mixed sector

Agriculture

About 30% of the workforce was usually involved in agriculture.

[] The main crops include wheat, barley, rice, vegetables, cotton, and dates (Iraq is one of the world's largest producers).

[] Iraqis also raise cattle and sheep.

Agriculture accounts for about 10% of the gross domestic product. Since the downfall of Saddam Hussein agricultural production has fallen off.

Foreign Dependence

Iraq depends very heavily on economic aid from both the West and Arab countries. The economic UN embargojust emphasized this. Both exports and imports were significantly reduced under the embargo, causing shortages and a rise in prices in Iraq.

Religious Division vice Iraqi Nationalism?

Two Sects

One of the important factors in understanding Iraq and its people is Islam.

[]The main Muslim sects of Islam, Sunnis, and Shias or Shi'ites, dominate the culture of Iraq. The majority in the country are the Shias, but the minority, the Sunnis, ran the country both before and during Saddam Hussein's regime. There was much discrimination against the Shias.

[]Most people feel that the Shias would have driven the Sunnis out of power if given the chance. (See appendix A for details on the differences between these two sects of Islam.)

Current Tension

One of the key issues today in post-Saddam Hussein Iraq remains the relationship between the majority Shias and the minority Sunnis.

[] The Sunnis are now apprehensive that with popular sovereignty, they will lose their political domination of the country.

[] How the majority Shias will treat the minority Sunnis is an important issue in post-Saddam Hussein Iraq.

Historical Division: One Side of the Coin

There was good historical reason for making the assumption that the Shias would have driven out the Sunnis.

[] For many years, the Arabian Sunnis from the areas around Baghdad, Mosul, and Ar Rutbah, the large cities of Iraq, dominated and ruled Iraq.

[]Saddam Hussein and his top deputies, all members of the ruling Baath Party, were also Sunnis. Sunnis had also held the top posts in the Iraqi security forces and had been the army corps' commanders.

Most of the Shias live in the south, the most depressed part of Iraq.

Continued on next page

Religious Division vice Iraqi Nationalism?, Continued

Possible Reverse Side: Nationalism

However, this historically based idea of "revenge" of the Iraqi Shias may not be as strong as assumed.

[]In 1982, in response to aggression by Saddam Hussein, Iran invaded Iraq. Iran, Iraq's neighbor, is primarily a Shia country.

[]The expectation of many was that the Shias of Iraq (75% of enlisted Iraqis were Shias) would join with their Shia brothers of Iran and revolt against the Sunnis of Iraq. No such general revolt occurred.

The Shias of Iraq continued to defend their country and the Sunni Baath regime of Saddem Hussein, even after the Iraqi army suffered major setbacks from the Iranian military.

Exploiting Ancient History

Perhaps what was at work here was the age-old antagonism between Arabs and Persians.

[] The Iraqis, both the Sunnis and the Shias, are an Arabic-speaking people.

[] The modern Iranians, while Shias in religion, are descendents of the Persians and speak Farsi, a language totally unrelated to Arabic. Between Arabs and Persians there has always been historically bitter hatred.

Saddam Hussein and his Baath Party capitalized on this ancient enmity. They described the war with Iran in terms of this hatred, characterizing the modern-day conflict as the "Battle of Qadisiyah," a medieval battle in which the Arabs, the true believers bearing Islam, defeated heathen Persia in A.D. 637.

Another "Division": Secular vice Religious

Islam vice Baath

Another cultural division, which has played a role in Iraq in more recent times, is the division between

[] the religious believers in the population, whether they are Shias or Sunnis,

and

[] the secularists or non-believers, who were primarily of the Baath Party.

Shia Opposition

Before the Baath Party, eventually headed by Saddam Hussein, came to power, Shias had made progress in education, business, and law even though they were under-represented in the government.

[]Later in the 1980s, Shias gained more representation in the government, proportional to their numbers in the population. For example, of the eight top Iraqi leaders who in early 1988 sat with Saddam Hussein on the Revolutionary Command Council—Iraq's highest governing body— three were Shias.

[]Saddam Hussein further suppressed the Shias so that they had no controlling voice in the Iraqi government.

The question remains if there will be a Shia "revival" in the "new" Iraq.

Tribes and Villages

Tribalism

As with many Middle Eastern countries, primary relationships play a key role in an Iraqi's life. Most family-oriented relations for an Iraqi are rooted in the age-old traditions of tribalism.

This tribalism is based on nomadism, but has been restructured to accommodate the sedentary environment in cities and towns and the limited territory of the modem age, e.g., land ownership.

Tribal Responsibility

The primary kinship unit is a tribe or *shaykh,* with the lineage through several generations from the father's side of the family.

[] This patrilineal (descent on the father's side of the family) kinship group is responsible for the extended family in feuds, arguments, and "wars."

[] It controls marriage and jointly controls parts of the "tribal" land. Mutual assistance is the key to a successful shaykh economy.

[] The "elders of the tribe" share authority.

[] A shaykh, which has existed through many generations, is usually very centralized and has a developed social hierarchy.

Clans

Within the tribe, the primary family unit is the clan, which is made up of two or more lineage groups. Clans are usually part of a particular tribe defined by blood or by adoption.

[]The clan supports its members in any dispute with other clans of the same tribe or shaykh.

[]Lands belonging to clans of a particular tribe are usually adjacent to one another.

Villages

The most widespread social living unit, the village, is usually a mixture of tribal folk, local tradesmen, and government employees. The line between tribal people and village dwellers used to be very distinct. As the government extended social services to the village, the number and influence of non-tribal people grew.

Continued on next page

Tribes and Villages, continued

Role Changing

Government and service people before and during Saddam Hussein's regime took over the role that tribal leaders and village people used to play. For example, a government school might be set up and competed with a village religious school. Government cooperatives might also work as the "middleman" selling farmer's goods, instead of the farmers doing this themselves.

[] In the 1950's most government employees came from Baghdad. With their citified ways, these Sunni bureaucrats often provoked antagonism among the local Shia villagers.

[] In the 1980's before the war with Iran, government representatives began to be drawn from the local village population where they were serving.

As you can see, the division between Shia and Sunni also plays a part in the Iraqi bureaucracy.

Iraqi History: A Rich and Ancient Past

Introduction

To get a better picture of the current political situation in Iraq, you will cover a bit about the history of the country. This brief historical survey is outlined as follows:

- "Ancient" Iraq
- Importance of Islam
- British Influence
- Background of the Rise of Saddam Hussein

Ancient Iraq: First Civilization

The area that modern Iraq covers is almost the same as that part of the ancient world called Mesopotamia (from the Greek, meaning "between two rivers," in this case the Tigris and the Euphrates). Another name for this region is the Fertile Crescent. Historians often refer to it as the cradle of civilization.

Like Egypt, China, and India, Mesopotamia was one of the ancient centers of civilization. In 4000 B.C., the Sumerians dominated the area.

- These ancient Sumerians built irrigation systems to harness the waters of the Tigris and Euphrates and developed the first cereal agriculture.

- They also created the earliest writing system, called cuneiform, by stamping clay tablets with a reed stylus.

- Banking also originated in ancient Mesopotamia.

Hammurabi

- The Sumerians were overrun and conquered by a number of Semitic-speaking tribes. (The Semitic languages include Arabic, Hebrew and a number of ancient, extinct languages.) Thus the Arab-related population became a force in the region early on.

- There was also much bitter infighting between the various tribes indigenous to the area. Finally around 1700 B.C., the region was united under King Hammurabi into one political unit called Babylon or Babylonia.

Continued on next page

Iraqi History: A Rich and Ancient Past, continued

Hammurabi's Dynasty: Achievements

King Hammurabi conquered territory and extended the borders of Babylon. He is remembered for his body of laws, Hammurabi's Code. Its common theme is that the punishment should fit the crime: "An eye for an eye, a tooth for a tooth."

Hammurabi's son, Nebuchadnezzar, designed and had constructed the ancient architectural wonder, the Hanging Gardens of Babylon, about 50 kilometers south of present-day Baghdad.

Islam: The Crown and Thorn in Iraq's History

Introduction

While ancient civilization is an important factor in Iraq's past, the conquest and establishment of Islam in the region is **the** historical and religious event that modern-day Iraqis are most proud of.

In A.D. 634, Arab Muslims reached the Euphrates delta. With the fervor of their Islamic jihad, *(Jihad fl Sabeel lillah)* they conquered and converted the Persians there, eventually establishing the Arabic dynasty of the Umayyads in Iraq.

The Caliphate

To understand the importance of Iraq in the history of Islam and the Middle East, you need to know about the caliphate.

[] Early on in the history of Islam after the death of Muhammad, Muslims chose a leader and established a "dynasty" for controlling all Muslims worldwide.

[]The caliph was the leader or head of this dynasty, known at the caliphate, which was the administrative, religious, and legal "directorate" of Islam. Later when the Ottoman Turks conquered Iraq and incorporated it into their empire, the caliphate was moved to Turkey.

[]The Republic of Turkey abolished the caliphate in the 1920's, but the caliphate as an historical tradition in Islam is very important for modern-day Iraqis.

The Abbasid Caliphate and Baghdad

Islamic culture and civilization in Iraq and the surrounding Middle East region reached its height under the Arabic dynasty of the Abbacies, who establish the Abbasid caliphate, a successor dynasty to the Umayyads, in Iraq in 750 AD.

The Abbasid caliphate, in what is present-day Iraq, founded the city of Baghdad, which became one of the cultural and economic centers of the Middle Ages. Baghdad was:

[] The intellectual center of the world at that time and the cultural capital of the Islamic world.

[] A power center where Arabic and Persian culture, intellect, and learning merged to create what all Muslims consider the apogee (high point) of Islam.

Safavids

Two groups vied for domination in the Iraqi region for the next two centuries. The Safavids were the Persian dynasty that ruled Persia (Iran), Iraq's neighbor. The Safavids:

[]Were Shias (Shi'ites) (remember, Shias are one of the two important divisions of Islam).
[]Wanted to extend Shia power throughout the Middle East.

The Safavids were very involved in attempting to conquer Iraq in the 16th and 17th centuries and in establishing Shia as the dominant sect of Islam in the region.

Ottomans

The Ottomans, a newly-arrived Turkish tribe, fast gaining ascendancy in Anatolia (modern Turkey) and Asia Minor:

[]Were Sunnis (the competing sect of Islam).
[]Did not want the Shias, led by the Safavids, to dominate Islam.
[] Sought to counteract the Shia aggression of the Safavids.

For the next two hundred year, the Persians (Iranians) and the Ottoman Turks conquered and re-conquered various parts of the Iraqi region. The Ottoman Turks eventually dominated Iraq and incorporated it into their empire.

(Good information on the early history of Iraq can be found at the web site http://home.achilles.net/~sal/iraq_history.html .)

Continued on next page

Iraqi Nationalism

One of the important legacies of the Ottoman domination of the Iraqis was the beginning of Iraqi nationalism.

- []Copying a group of Ottoman Turkish nationalists called the Young Turks, Iraqi intellectuals founded an Iraqi nationalist society called *Al Ahd* (the Covenant) in the early 1900s.

- []The aim of this group was to gain more autonomy for Iraqis from the Ottomans.

Offshoots of ***Al Ahd*** formed "cells" among Iraqi officers serving in the Ottoman army and in larger towns such as Mosul and Baghdad.

(For more information on this period see the following web site: http://mcmorv.loc.gov/cgi-bin/qucrv/r?frd/cstdv:@ficld(DOCID+iq0018))

British Influence

Post World War I

The British invaded Iraq as part of their campaign against the Ottoman Empire during World War I. After the war, the Treaty of Sevres gave Great Britain control of Iraq as a mandate of the League of Nations.

[]In 1921 the country was given a monarchy headed by King Faisal I.

[]By 1926, an Iraqi parliament was governing the country although the British, under a 1924 treaty, had a veto over legislation.

[]The British mandate ended in 1932 when Iraq was admitted to the League of Nations.

[] Iraqi oil exports began in 1934.

[] During the period 1936-41, a time of instability in the country, Iraq experienced seven military coups.

(For a fascinating narrative on how the British set up the borders of Saudi Arabia, Kuwait, and Iraq in 1922, go to the web site http://www.nvbooks.com/articles/3032.)

World War II

Initially, Iraq sided with Germany and Italy in World War II, but after the pro-Axis Iraqi leader, al-Gaylani, was defeated by the British in early 1943, the country joined the Allies and declared war on Germany.

Rise of Saddam Hussein: Background

Key Factors

To understand the rise of Saddam Hussein to power in Iraq, you need to know about certain key facts in Iraqi political history:

- []Constant conflict including coup and counter-coup has been a common feature of Iraqi political life for many years.
- []The relationship of Iraq with its neighbors, particularly Iran, played a role in influencing Saddam Hussein.
- [] The growth of the Baath Socialist Party was important too.

We will consider this context in discussing Saddam Hussein's emergence as Iraq's leader.

(This is an excellent article on the family "clan" of Saddam Hussein: http://ww w.mcforum.ori>/articlc/273 .)

Continual Conflict

The period of the fifties, sixties, and seventies was one of continual conflict in Iraq:

- []In the late 1950s the monarchy was overthrown.
- []During the 1960s the Baath Socialist Party under Hasan al-Bakr came to power in the country. Coup and counter coup with espionage trials were common during this time.
- []The Baath-led government was overthrown in a coup led by President Abdul Salam Mohammad Arif, who was succeeded by his brother, General Abdul Rahman Arif.
- []In July 1968, the Baath Party regained power in a coup under Ahmed Hasan al-Bakr, who then became president of the country.

Continued on next page

Rise of Saddam Hussein: Background, Continued

Iraqi Foreign Relations

To understand some of the reasons for Saddam Hussein's aggression toward his neighbors, you need to know a bit about Iraq's relations with neighboring countries before Saddam's rise to power in the country.

[]Iraq participated in the Arab-Israel War of 1973 and in the oil boycott against nations supporting Israel.

[]In early 1974, continual border clashes with Iran erupted in an armed conflict along the entire Iraq-Iran border.

[]In 1975 the Kurds, supported by the Iranians, again revolted in northern Iraq, fighting for their independence. Iraq bombed parts of Iran, which created more tension between the two countries.

[] In March 1975, Iraq and Iran signed the Algiers Agreement, which defined the boundary of the Shatt el-Arab waterway. Iran then withdrew support from the Kurds, who suffered enormously with intense Iraqi bombings.

[]The Islamic revolution in neighboring Iran was recognized as a threat to secular Baath party control of Iraq and caused great anxiety in Iraq.

As you can see, military aggression has played a role in Iraq's relations with its neighbors!

Detailed information on this period and on the rise of Saddam Hussein is at

http://mcmorv.loc. gov/cgi-bin/qucr\'/r?l"rd/cstdv:@field(DOCID+iq00231

http://mcmorv.loc. gov/cgi-bin/qucr\'/r?l"rd/cstdv:@ficld(DOCID+iq00221.

Rise of Saddam Hussein: Baath Party

Early History

The Iraqi Baath Party was founded in the 1940s by two Syrian students.

- The word *Ba'ath* or *Baath* means resurrection in Arabic; hence, the party's goals were to "resurrect" the Arabs of Iraq as an ethnic group through socialism, freedom, and unity. This was supposed to occur in Iraq the way it was happening in neighboring Syria.

- The party grew quickly in the early 1950s. The party had difficulty after an assassination attempt in the late fifties.

One Party State: Saadi

Saadi, one of the chief leaders of the party, was the Baath political leader who consolidated the party's power. Although the Baathists established the National Council of Revolutionary Command, which appointed the prime minister and the president, Saadi was the real power behind the government. In other words, the party controlled the state.

What the party needed in its role as the nation's leader was cohesion. Saadi gave it this cohesion. He established a one-party state that tolerated little opposition.

Coup and Counter Coup

The Baath-led government was overthrown in late 1963, only to regain power in 1968.

The Baath Party of 1968 was much better organized and was a more cohesive unit than in 1963. The party also was aided by its reorganization.

- The party gained broader support based on its newly established militia and intelligence system.

- It also had local party branches that supported the central Baath Party.

Continued on next page

Rise of Saddam Hussein: Baath Party, Continued

The Tikrit Connection

Tribal tics bound ruling members of the Baath Party together.

[] Most of them were Tikritis, Sunni Iraqis from Tikrit, a town in northwestern Iraq.

[]Some of the parly members were also related to the new president, al Bakr. Besides the prime minister, who was Tikriti, Saddam Hussein himself was also related to al Bakr and from a town near Tikrit.

The one-party system was thus reinforced by this coterie of Tikritis, who wielded the political power in the country.

More information on Saddam Hussein can be found at:

http://www.bartlcbv.com/65/hu/HusscinS.html

http://homc.achillcs.net/~sal/iraq_history.html

Saddam Hussein, Former Leader of Iraq

Early Years

Saddam Hussein had been a long-standing member of the Baath Party of Iraq. He fled Iraq in 1959 after taking part in an assassination attempt against then Prime Minister Kaseem. He then lived in Egypt and attended law school there.

- Saddam Hussein returned to Iraq in 1963 when the Baathists came back to power for a short time.
- Later he played a leading role in the 1968 revolution that gave the Baath Party control of the country, supporting Ahmed Hasan al-Bakr, who became the Baath chairman of the Revolutionary Command Council and the President of Iraq.

The Baathist Tikrit Team

President al Bakr and Saddam Hussein, both Tikritis, dominated the Baath Party, which returned to power in 1968.

- Al Bakr had espoused pan-Arab causes and gave the party its legitimacy. He also ensured support from the army among Baathist and non-Baathist officers.
- Saddam Hussein was the politician. His expertise was in organizing secret opposition operations. He knew how to outmaneuver and eliminate political opponents when necessary.

As time went on, it became clear that Saddam Hussein was the moving force behind the party, the political aficionado who worked behind the scene. He personally

- Directed Iraqi efforts to deal with the Kurdish problem
- Organized the party's infrastructure

Continued on next page

Saddam Hussein, Former Leader of Iraq, Continued

Saddam Hussein — alone

In the mid-seventies, Saddam Hussein began to consolidate his political power.

[]With President al Bakr in failing health, most of the ministers, who should have been reporting to the president, reported directly to Saddam Hussein.

[]Saddam Hussein showed his modus operandi: He did not share power and considered the cabinet and the council rubber stamps for his will.

In July 1979, President al Bakr resigned, and Saddam Hussein officially became President of the Republic, Secretary General of the Iraqi Baath Party Regional Command, Chairman of the Regional Command Council, and Head of the Iraqi Armed Forces.

Saddam Hussein's Policies

Introduction

Purpose

Rather than "narrate" the history of Iraq under Saddam Hussein's dictatorship, we will choose key topics that show his policies. The topics discussed show his domination and oppression.

Selected Topics

The topics covered in this section are shown in the table below.

Kurdish Oppression

Early Oppression and Resistance

Kurdish history, whether in Turkey, Iran or Iraq, has been a tradition of oppression with the aim of controlling the Kurdish population. This has resulted in strong military resistance from the Kurdish side. The Iraqi government has characterized the Kurdish response as rebellion and revolt. In Iraq the oppression and its response is outlined thus:

[] In 1958 after the overthrow of the Iraqi monarchy and the establishment of the Baathist-led government, the Kurds hoped for more autonomy in administration. The new Iraqi government failed to follow through.

[] During the 1960s, Iraqi Kurds under the leadership of Mustafa al-Barzani waged a lengthy military campaign with the aim of establishing a unified and autonomous Kurdistan in Iraq.

1970's

In 1970 the central government promised local self-rule to the Kurds. Their capital was to be Erbil. The Kurds rejected the proposal for two reasons:

[]The proposed autonomy did not really give Kurds authority. The central government still retained control.

[]As part of the newly proposed Kurdish autonomous region, the Kurds were not granted control of Kirkuk, an important city in the Kurdish area and a key oil-producing center.

Heavy fighting between the central government and Kurdish forces erupted when the central government tried to force limited autonomy on the Kurds in 1974.

Continued on next page

Saddam Hussein's "Campaign"

In 1979, the Baathist-lcd government of the new Republic of Iraq launched a campaign of terror and murder against the Kurds. It also carried out a program of systematic assassination of Kurdish leaders.

Saddam Hussein's reign of terror continued for eight years, culminating in 1987-1988 with the Anfal Campaign.

[] His forces tortured and killed Kurds.

[] He destroyed many Kurdish villages, forcing the surviving villagers to live in special zones and camps where he could control them.

Finally, Saddam Hussein ordered the use of poison gas attacks on Kurdish villages and the arrest and execution of Kurdish males in a final effort to root out all resistance. The result that year was the murder of 200,000 people.

Post-1991 Persian Gulf War

In 1991 after the end of the Persian Gulf War, Iraqi forces again crushed another Kurdish uprising. Fearing gas attacks again, nearly 500,000 refugee Kurds attempted to flee to Turkey, and one million fled over the border into Iran.

Later when an autonomous U.N.-protected region for Iraqi Kurds was established in northern Iraq, thousand of Kurds returned to and resettled in Iraq. A general election was held in the autonomous region.

Kurdish "Politics"

The Kurds themselves have internal problems. They are split into two opposing factions, the Kurdistan Democratic Party and the Patriotic Union of Kurdistan. In 1999 these two groups ended hostilities.

Internal Control

Cult Atmosphere

Saddam Hussein created a personality cult that fed on his will, totally. He effectively silenced all opposition to his regime in Iraq. For over 20 years, his style of oppression ruled Iraq, silencing a people that had once been a rich source of thought and culture in the Middle East.

Suppression of Human Rights

[]Saddam Hussein did not allow his people to vote.

[]Iraqis had no freedom of expression. The daily Iraqi newspaper was owned bv his son.

[]There was no freedom of association and movement in the country. Iraqis, allegedly, could only assemble to show their support for Saddam Hussein and could only vote for him in elections. Iraqis were not free to leave the country.

[]Saddam Hussein established a large network of spies, through blackmail and fear. In some instances, members of a family were so terrorized that they reported on their relatives.

Continued on next page

Internal Control, Continued

Regime of Systematic Torture and Terror

The following allegations have been made against the regime of Saddam Hussein:

[]Collective torture — Saddam Hussein used collective torture to suppress dissent and to instill fear in anyone considering opposing his will. Whole families or ethnic groups (the Kurds were a good example) were tortured for the actions of one dissident.

[]Public torture and humiliation — Saddam Hussein's forces respected no person's dignity. Women were allegedly raped and tortured in front of their families, many were videotaped in order to blackmail their families. Suspect citizens were cruelly dismembered; their families had to pay for the instruments of death and then display the body as a warning to others who might be considering expressing an opinion.

[]Branding and Amputation — Hospitals were known to brand dissidents or to amputate their limbs. In 1994, Saddam Hussein issued a number of decrees establishing cruel and inhuman penalties such as amputations. Military deserters were punished by amputation.

[]Chemical weapons — Saddam Hussein also used chemical weapons against his own people. He decimated more than 60 villages and murdered 30,000+ Iraqis with poison gas.

[]Summary executions — There were numerous ways that Saddam Hussein's forces did this. One "efficient" method was to line up all the males of a village and shoot them one at a time. But much of the time, a slow, excruciating death was preferred. Political prisoners were given slow-acting thallium, a poison that "permeates" the human body and takes several days to bring death. In many instances, the bodies were buried in unmarked sites.

Continued on next page

Internal Control, Continued

Religious and Social Atrocities

Besides oppressing the Kurds, the largest ethnic minority in the country, Saddam Hussein tried to silence other ethnic and religious groups.

[] He waged terror and oppression against religious believers of Iraq.

[] Through murder, torture, and limitations on their religious activities, he suppressed many clerics and scholars in the country, particularly the Shia Muslims.

Anyone whom Saddam Hussein feared would oppose his will was subjected to torture and arbitrary death. This began almost immediately after he became president in July 1979. Claiming "traitors" in the Iraqi National Assembly, he ordered his security forces to publicly and forcibly imprison and eventually kill several leading members of the Iraqi National Assembly.

Children: Pawns of Saddam Hussein

Children, allegedly, were notjust careless "by-stander" victims in Saddam Hussein's reign of terror and suppression.

[] Pharmaceutical supplies for sick children were supposedly exported for resale overseas. Medical supplies were often delayed because the regime demanded bribes from the suppliers. Poor health care for children led to the reemergence of diseases such as polio and cholera in Iraq.

[] Saddam Hussein's forces supposedly abducted and held minority children hostage to force their families to relocate in particular regions, especially those families near oil-producing areas.

[]There were also weapons training courses for teenagers, subjecting them to 14 hours of physical training and psychological stress each day.

[]In many cases, children were allegedly forced to become spies, reporting on their parents' and relatives' conversations and activities to Saddam Hussein's internal security organization.

(See the site http://www.state.gOv/g/drl/rls/15996.htm for details on how Saddam Hussein is silencing and controls his own people.)

International Response to Saddam Hussein's Oppression

United Nations Resolutions

In Security Council Resolution 677, November 28, 1990, the United Nations condemned Saddam Hussein for forcibly resettling various groups in order to alter the population in different areas of Iraq.

In Security Council Resolution 688, 5 April 1991, the United Nations condemned the repression of Iraqi civilians.

United Nations Reports

After a 1999 trip to Iraq, the United Nations special representative on human rights reported he had personally received reports and testimonies of

[]People who showed him their scars from government torture

[]Women whose family members had been abducted by Iraqi authorities and then had disappeared.

He felt there was a prevailing regime of systematic human rights violations in the country.

In the United Nations Secretary General's Report of 2001, the U.N. special representative stated,

> "TTie *mere suggestion that someone is not a supporter of the President* [Saddam Hussein/ *carries the prospect of the death penally."*

Iran-Iraq War, External Domination

Background

[]As discussed earlier, the antagonism between Iran and Iraq is age-old. It is based partly on the clash of two different cultures: one, Arab-based; the other, Persian.

[]Add to this, the religious competition between the two prominent Islamic sects, the Sunnis, who control Iraq, and the Shias, who control Iran, and there is bound to be constant tension between these two giants of Southwest Asia.

Political Aims

Saddam Hussein's aim at domination focused on this constant friction point between Iraq and Iran. His foreign policy emphasized

[] Strengthening Iraq's role in world oil production
[] Gaining the leading role in the Arab world and the Middle East

Accomplishing these two aims had to come at the expense of Iran, which under both the former Shah and the later Islamic religious regime of Ayatollah Khomeini had also sought to expand its influence in these two areas.

Causes of the War

[] In 1980, Saddam Hussein escalated a dispute with Iran over control of the Shatt al Arab waterway by attacking Iran.

[] Besides the territorial claim over the waterway, Saddam Hussein also maintained that Iran had been carrying out artillery attacks on Iraqi territory.

Continued on next page

Iran-Iraq War, External Domination, continued

Territorial Aims

Saddam Hussein wanted to

[] Reclaim the Shatt al Arab waterway for Iraq
[] Seize the western region of Khuzestan in Iran, an area of extensive oil fields

Western Involvement

[] France, Germany, and the Soviet Union were the main arms suppliers to **Iraq.** United States and Great Britain also quietly provided arms and related military assistance.

[] Iran received funding from some Arab countries.

Initial Success; Later Defeat

The Iraqis were successful at first in their aggression. They captured the Iranian port of Khorramshahr by late 1980; however, Iranian resistance strengthened so that Iraq had to withdraw in 1982.

Ayatollah Khomeini, the leader of Iran, declared that his country would only stop fighting when Saddam Hussein's government had been destroyed. Iran then began a series of successful assaults that caused Saddam Hussein to use poison gas against Iranian troops.

[]Iran captured and occupied some oil-rich islands from Iraq in 1984.
[]Both countries began to attack each other's capital cities in 1985.
[]Iran captured part of Iraq's Fao peninsula in 1986.

Ending the War

[]After Iran began to attack Kuwaiti oil tankers in the Persian Gulf, the United States and some European nations got involved in the struggle.

[]Attacking a neutral nation like Kuwait caused problems for Iran. The country was not able to buy enough weapons to continue the war.

Finally, in July 1988, Iran accepted a United Nations cease-fire.

Summary

Saddam Hussein's principal aim in the Iran-Iraq War was to extend his power in the Middle East. He wanted to dominate the area economically in terms of oil production and militarily by acquiring Iranian territory. He didn't succeed.

Persian Gulf War

Causes of the Kuwaiti Invasion

Saddam Hussein's invasion of Kuwait, which was the ultimate reason for the Persian Gulf War against Iraq, was the result of a territorial dispute. Saddam Hussein accused Kuwait of the following:

[] Violating the Iraqi border to get oil resources from the disputed Rumaila oil field

[] Flooding the world market with Kuwaiti oil, causing a decrease in oil prices

Failed Negotiations

[] After a number of failed meetings between Kuwait and Iraq, sponsored by Arab mediators

[] After assurances of the United States' noninvolvement in the dispute

Saddam Hussein's forces invaded and occupied Kuwait on August 2, 1990, later annexing the country as the nineteenth province of Iraq.

United Nations Involvement

The United Nations Security Council and the Arab League condemned the invasion. Acting under Articles 39 and 40 of the United Nations Charter, the United Nations passed Resolution 660, which

[]Condemned the Iraqi invasion of Kuwait

[]Demanded immediate withdrawal of Iraqi forces from Kuwait.

The Embargo and Further UN Resolutions

Four days later, the United Nations passed an economic embargo that prohibited nearly all trade with Iraq. There followed a series of United Nations resolutions including:

[]Further condemning the Iraqi invasion

[]Demanding again the withdrawal of Iraqi forces

[] Expressing sympathy for and grave concern over the treatment of Kuwaiti nationals in occupied Kuwait

[]Stressing the importance of the humane treatment of Kuwaitis under terms of the Geneva Convention

[] Laying out specifics of the international embargo against Iraq

Continued on next page

Persian Gulf War, Continued

Resolution 678

This series of U.N. resolutions culminated in United Nations Security Council Resolution 678: It allowed U.N. member countries to use all necessary means, including military action, against Iraqi forces occupying Kuwait and demanded total withdrawal of Iraqi forces by January 15, 1991.

United Nations Coalition War

Operation *Desert Shield,* preparations and the build-up for the eventual invasion of Iraq, began in late 1990.

On January 17, 1991, the U.N. allied coalition of 28 countries, led by the United States, launched aerial attacks on Baghdad in Operation *Desert Storm,* popularly known as the Gulf War.

[]The war lasted six weeks. Over 140,000 tons of munitions were unleashed over Iraq. About 100,000 Iraqi soldiers were killed. Bombing destroyed the Iraqi infrastructure including the oil refinery system. All services were totally disrupted.

[]On February 24, 1991, the U.S.-led coalition forces entered Iraq with ground forces as part of the plan for liberating Kuwait, but did not penetrate as far as the Iraqi capital Baghdad.

Conclusion to Hostilities

[] The United States announced a cease-fire on February 28, 1991.

[] In April 1991, Iraq agreed to a permanent cease-fire with strict conditions.

[]Saddam Hussein still remained in power as the head of the Iraqi government.

This site discusses the reconstruction of Kuwait after Saddam Hussein's defeat and the withdrawal of Iraqi forces: http://www.mepc.org/public asp/journal vol7/0006 tetreault.asp .

Defiance of United Nations: Restrictions on Iraq

Continued Embargo

United Nations Resolution 687 began the cease-fire, created the United Nations Special Commission on Weapons and continued the embargo tying it to Iraqi weapons. The sanctions sought to

[]Eliminate Iraq's weapons of mass destruction and delivery systems

[]Return to Kuwait all prisoners of war and property taken during the Gulf War

[]Establish the principle of Iraqi compensation for war damages

Further the sanctions

[] Insisted that Iraq honor its international debts

[]Demanded that Iraq stop all terrorism

A Goods Review List (GRL) listed all those items under embargo.

More Minority Persecution

One of the key problems in Iraq after the United Nations cease-fire was Saddam Hussein's continued oppression of the Kurds of northern Iraq. Iraqi aircraft and troops were sent there to regain control of the cities that the Kurdish resistance groups had taken during the Gulf War.

Fearing a repeat of the gas attacks of 1988, an exodus of Iraqi Kurdish refugees (about 2.5 million in total) massed on the Turkish and Iranian borders. In the south, Saddam Hussein's forces attempted to control the marsh Arabs, many of whom are Shias.

[] United Nations Resolution 688 condemned this continued repression of the Iraqi civilian population.

[] Operation *Provide Comfort* established a United Nations protected autonomous region in most of the Kurdish areas of Iraq.

Continued on next page

Exclusion Zones

To prevent the genocide of the Kurds in the north and the marsh Arabs in the south, the United States, together with France and Great Britain, established air exclusion zones or no fly zones north of the 36th parallel and south of the 32nd parallel, areas prohibited to Iraqi forces.

- The Iraqi air force was not allowed into these exclusion zones. These no-fly zones allowed Kurdish refugees to return home and live without fear of Iraqi reprisal in their autonomous region as allied planes continually surveyed and guarded these exclusion regions.

- As of 1 October 2000, it was reported that in 16,000 sorties since the beginning of 1997, air force pilots had launched more than 1,000 bombs and missiles against 250 targets in northern Iraq in protecting that no fly zone. (See Thomas E. Ricks, "Containing Iraq: A Forgotten War," *Washington Post,* October 25, 2000.)

There was also a no drive zone in southern Iraq to prevent Saddam Hussein from again massing forces on the Kuwait-Iraq border.

Continued Incidents

During the Clinton administration, Iraq attempted to assassinate former President George Bush while he was visiting Kuwait. The United States response was to fire cruise missiles at Iraqi intelligence headquarters in Baghdad.

In October 1994, Saddam Hussein's republican guard moved toward Kuwait. The United States responded with troop deployments to the Persian Gulf. Saddam Hussein backed down and agreed to recognize the existence and borders of Kuwait.

Sanctions Modifications

The original economic sanctions imposed on Iraq had been total. Because of humanitarian needs of the Iraqi civilian population, the United Nations modified the sanctions by establishing the Oil for Food Program in 1995 in United Nations Security Council Resolution 986.

- This resolution allowed Iraq to sell a billion dollars of oil to purchase humanitarian supplies, primarily food and medicine, and supplies necessary for the civilian infrastructure of the country.

- The first shipments under this program began arriving in Iraq in 1997.

Continued on next page

Defiance of United Nations: Restrictions on Iraq, Continued

Sanctions Modifications, (continued)

[] Under this program, Iraq could sell oil for humanitarian needs.

[] The United Nations controlled all revenues from such oil sales, and the contracts to implement this program were subject to United Nations oversight.

[] The program was extended and modified through the years.

A Slow Process

The process of "food" for oil was slow.

[]Some funds from oil sales might have been diverted to purchase non-humanitarian supplies.

[]Contract reviews were tedious.

Critics of the program called for revised sanctions that were more targeted and credible.

The U.S. Department of State web site has a myth/facts sheet about the effect of sanctions on Iraq at http://www.state.go\7p/nea/rls/01fs/3935.htm .

http://homc.achillcs.net/~sal/un-ros.html has a list of the United Nations resolutions on sanctions.

Arms Inspections and Disarmament

Purpose

[] The principal purpose of United Nations Resolution 687 was to eliminate Iraq's capability to inflict damage and make war on its neighbors. Essentially the resolution required the elimination of Iraq's weapons of mass destruction and delivery systems.

[] The resolution also charged Iraq to "forego the future development or acquisition of weapons of mass destruction (WMD)."

Definition

Four categories of WMD, for declaring, destroying, or rendering harmless, were defined in the resolution:

[]Nuclear weapons
[] Chemical weapons
[]Biological weapons
[]Missile delivery systems

Implementation

To ensure that Iraq followed the resolution by declaring, destroying or rendering harmless its WMD, the United Nations:

[]Established The United Nations Special Committee (UNSCOM).

[]Charged UNSCOM and the International Atomic Energy Agency (IAEA) to inspect Iraq to ensure its compliance with the United Nations resolution on declaring, destroying, and rendering harmless Iraq's WMD.

Inspections, 1991-1998

Iraqi Obstructionism

From the start of the UNSCOM inspections, as they were called, Saddam Hussein had been determined to keep a sizeable part of his WMD arsenal.

[]This led to constant dissembling and obstruction on the part of the Iraqi government during UNSCOM inspections.

[]Saddam Hussein's security force coordinated a campaign of concealment and deception, including hiding documents related to WMD and materiel. This made it extremely difficult to resolve, let alone inspect, issues and materiel related to Iraq's WMD program.

[] Iraqi officials outwardly maintained a facade of cooperation, while at the same time delaying or denying UNSCOM inspectors access to facilities, documents, and personnel in a purposed effort to deny information to UNSCOM on Iraq's WMD program.

UnitedNations Response

Prompted by this Iraqi obstructionism, the United Nations Security Council passed several resolutions demanding that the Iraqi government

[] Cooperate with UNSCOM and IAEA in the inspection process

[]Provide UNSCOM and IAEA with " immediate and unrestricted access to any site they wished to inspect"

The Iraqi Arsenal

The coalition military strikes during the 1991 Persian Gulf War and UNSCOM inspections through 1998 destroyed most of Iraq's prohibited ballistic missiles and some Gulf war-era chemical and biological munitions. Iraq still had

[] A small force of extended-range Scud missiles
[] Some pre-chemical weapons
[] Stocks from which biological weapons could be produced
[] Munitions suitable for chemical and biological agents

Continued on next page

Inspections, 1991-1998, continued

The Iraqi Arsenal, (continued)

Following the Gulf War, Iraq may have also

[]Preserved and enhanced the infrastructure and expertise necessary for WMD production

[]Maintained a stockpile of WMD and increased its size and sophistication in some areas

Hiatus in United Nations Weapons Inspections

In December 1998, the Iraqi government refused to continue working with UNSCOM,

[] Onsite inspections by UNSCOM were prohibited.

[]Technical monitoring systems installed by the United Nations at known and suspected WMD and missile facilities did not operate.

[]Iraq prohibited Security Council-mandated monitoring over-flights of Iraqi facilities by United Nations aircraft and helicopters.

[]Iraq had also stopped most IAEA inspections since 1998.

After the United Nations staff had been evacuated from Baghdad, the United States and Great Britain launched a bombing campaign, Operation *Desert Fox,* to destroy Iraq's nuclear, chemical, and biological weapons programs.

http://www.cia.gov/cia/publications/iraq_wmd/Iraq_Oct_2002.htm is a detailed discussion of Iraq's WMD and the UNSCOM effort, particularly through 1998.

Important Post-1998 Events

Terrorism: The Focus

The terrorist attacks on the World Trade Center in Manhattan and the Pentagon, with its connection to Al Qaida-sponsored terrorism, brought Saddam Hussein and Iraqi armaments to the forefront of the Bush administration's policy on terrorism.

Assertions by President Bush and his administration now focused on Saddam Hussein as the primary terrorist security risk to the United States and the world, emphasizing the following:

[]A connection existed between Saddam Hussein and terrorism by his giving "aid and comfort" and training to such terrorists.

[]An axis of evil, including Saddam Hussein, is a concerted effort at state-sponsored terrorism.

[] Saddam Hussein had never met his disarmament obligations under numerous United Nations resolutions and implementation inspections.

[]There was a dangerous possibility that Saddam Hussein might supply terrorists with WMD, particularly biological weapons, such as smallpox.

Heightened U.S./U.N. Actions

In September 2002 President Bush urged the United Nations to encourage Saddam Hussein to comply with United Nations resolutions or "actions will be unavoidable."

[] Bush said that Saddam Hussein had repeatedly violated 16 United Nations Security Council resolutions, which included a call for Iraq to *"disarm its chemical, biological, and nuclear weapons programs."*

[] Iraqi officials rejected Bush's assertions.

In response to the United States' heightened effort and possible direct action against Iraq, the United Nations passed resolution 1441. (See the next section for provisions of this resolution.) In mid-November, Iraq accepted the return of United Nations weapons inspectors under resolution 1441.

(See the UN site, httD://www.un.or2/aDDs/ncws/stor\'Ar.asD?NcwsID=5344&Cr=iraa&Crl= for information on this acceptance.)

Continued on next page

Important Post-1998 Events, continued

United Nations Security Council Resolution 1441: Important Features

The key provisions of Resolution 1441 were that Iraq must:

[]Submit to the United Nations Monitoring, Verification and Inspection Commission (UNMOVIC), the IAEA, and the Security Council a complete accounting of all its nuclear, biological, and chemical programs, all ballistics missiles and other delivery systems, with a precise listing of all weapons locations, stocks of agents, and all production facilities.

[]Provide UNMOVIC and IAEA with "unimpeded, unconditional, and unrestricted access" to all types of facilities and transport which the two agencies may want to inspect.

[]Allow unrestricted interviews to all personnel that UNMOVIC and IAEA may want to talk with, either inside or outside Iraq. The families of interviewees may also travel with the interviewees. Interviews may occur without the presence of Iraqi government observers.

If Iraq failed to comply with this resolution, it would be in violation of the resolution, which would be reported to the Security Council. The resolution said nothing about the "punishment" that Iraq would incur if it didn't fulfill this resolution.

A complete text of the resolution is at: http://www.un.org/Docs/journal/asp/ws.asp?m=S/RES/T441(2002) or at http://ods-dds-nv.un.org/doc/UNDOC/GEN/N02/682/26/PDF/N0268226.pdf?OpenElement

The Mandate

United Nations Security Council Resolution 1284 created the United Nations Monitoring, Verification and Inspection Commission (UNMOVIC), which replaced UNSCOM. The mandate of UNMOVIC was to

[]Continue with UNSCOM's efforts to disarm Iraq of its weapons of mass destruction

[]Check and monitor Iraq's compliance with its obligations not to reacquire the weapons prohibited by the United Nations Security Council

(The United Nations site http://www.un.org/apps/news/infocusRel.asp?infocusID=50&Bodv=Iraq&Bodyl=inspcct has information on the resolutions and the inspections.)

Iraq's Weapons Declaration

Compliance

Iraq allowed United Nations inspectors back into the country. Cooperation with U.N. personnel, however, was not full or complete.

- []The Iraqi government did not allow personnel associated with its various WMD programs or facilities, such as scientists and technicians, to talk freely with UNMOVIC inspectors. An Iraqi security official was invariably present.

- []Iraq submitted its required weapons declaration to the United Nations, but the report had irregularities and gaps, some of which are discussed below.

Biological Agents

In the declaration on its weapons of mass destruction to the United Nations, Iraq omitted a number of key biological agent items, including:

[] 2,160 kilograms of growth media for anthrax
[] 1,200 liters of botulism toxin
[] 5,500 liters of Clostridium

The United States asked why the Iraqi government ignored these agents in its declaration.

Ballistic Missiles

[] Iraq said it was manufacturing new energy fuels suited for a class of ballistic missile which it had not admitted it had previously possessed.

[] The United States government asked why Iraq was manufacturing fuel cells for missiles it sairl it Hirln't hnve

Continued on next page

Iraq's Weapons Declaration, continued

Chemical & Biological Weapons Munitions

In January 1999, the United Nations Special Commission reported that Iraq had **not** provided credible evidence that:

[]550 mustard gas-filled artillery shells and 400 biological weapon-capable aerial bombs had been lost or destroyed.

[]The Iraqi government had not adequately accounted for these chemical weapons in the declaration.

The U.S. government asked what the Iraqi government was concealing by not including these munitions in its declaration.

Mobile Biological Weapon Facilities

The Iraqi declaration provided no information about its mobile biological weapon agent facilities, i.e., biological labs in trucks. Instead it insisted that these were "refrigeration vehicles and food testing laboratories."

Again, the United States asked what was the Iraqi government concealing about these mobile facilities.

More information about the declaration is at
http: //www, state, gov/r/pa/prs/ps/2002/16118.htm

A New Coalition

The Smoking Gun

Many members of the United Nations were looking for direct evidence of Iraq's WMD before considering the possibility of any intervention in the country.

- []Russia, China, France, and Germany were against immediate military intervention in Iraq.
- []The reluctance of other United Nations members made it difficult for the United States to build an allied coalition for invading Iraq.

Beginning a Coalition

The United States was working on building a new coalition for the possible attack on Iraq. Many nations that participated in the 1991 Persian Gulf War were reluctant to join the coalition.

- [] France, a permanent member of the Security Council, had not ruled out military force against Iraq, but said that United Nations inspectors must be given the broadest opportunity to fully do theirjob; otherwise, France would veto any United Nations resolution authorizing force against Iraq.
- []Germany said it would not support a military campaign.
- []Popular opinion in much of the world was against military intervention without United Nations approval.

Turkey

The Republic of Turkey is an example of a reluctant coalition partner. Turkey is a non-Arabic secular state.

- []It had to be one of the key countries in any coalition strategy against Iraq because part of the Bush administration's military strategy was to attack Iraq from the north.
- []The northern prong of this possible invasion would come from Turkey.

Continued on next page

A New Coalition, Continued

Turkish Apprehension

[]Most Turks are Muslims like the Iraqis.

[]Additionally, public opinion in the country was not in favor of the war as many Turks felt that the war might engulf the whole Middle East and destabilize their country politically and economically.

[]There was also the fear that Saddam Hussein would retaliate directly against Turkey if the country were used as a base for invading Iraq.

[]At this point, the Turkish parliament voted against stationing United States and coalition troops that might transit the country.

The United States continued to work on an acceptable allied coalition strategy with nations such as Turkey.

Joining Up

The United States' staunchest ally in forming a coalition for the invasion of Iraq and in support of reconstructing post-war Iraq has been the United Kingdom. Prime Minister Tony Blair carried out shuttle diplomacy to persuade reluctant nations to join the U.S.-led coalition.

In late January 2003, Blair, together with Prime Minister Jose Maria Aznar of Spain, led a drive to persuade other European nations to join the coalition.

[]Six additional nations - Italy, Denmark, Portugal, Poland, Hungary, and the Czech Republic - signed on.

[]Most glaring was the absence of the two giants of Europe, Germany and France, which both had very strong reservations about military action against Iraq.

(See the site http://www.washingtonpost.com/wp-dvn/articles/A3685-2003Jan30.html for details on coalition building.)

Continued on next page

Making the Case against Saddam Hussein

On February 5, 2003, the U.S. Secretary of State, Colin Powell, presented the U.S. case against Iraq to the United Nations general assembly. Using satellite photographs and intercepted audio messages from the Iraqi military, the Secretary of State offered evidence that Iraq

[]Still had WMD
[]Was not cooperating with the U.N. inspectors
[]Had links to Al Qaida (Al Qaeda)

Military Intercepts

Powell played recordings of tèlephone messages between high-ranking Iraqi military officers discussing

[]"Forbidden ammo"
[]Concealment of modified vehicles from U.N. inspectors

Concealment

In maintaining that Iraq still concealed WMD, the Secretary of State discussed the following:

[]Iraq had bulldozed and concealed alleged chemical weapons at the Al Musayyib chemical complex in 2002 and had a series of cargo vehicles and decontamination vehicles moving around at the site.

[]He cited intelligence sources showing that the Iraqis had been dispersing rockets armed with alleged biological weapons in western Iraq.

[] Iraqi informants claimed that Iraq had 18 trucks that it used as mobile biological weapons labs.

Continued on next page

A New Coalition, Continued

Al Qaida Connection

Powell also maintained that

> *"Iraq harbors a deadly terrorist network headed by Abu Musab Zarqawi, an associate and collaborator of Osama bin Laden and his al Qaeda lieutenant."*

[] Zarqawi spent two months in Baghdad during May and June 2002 getting medical treatment.

[] Some members of his group were supposedly based in Baghdad.

Reaction

Reaction among the U.S. population to Powell's presentation was generally positive.

[]In the United Nations, member states, which had had reservations about military intervention in Iraq, felt that the UNMOVIC should still be given more time to carry out further inspections.

[]Because of member states opposition to military intervention and the threat of a veto by other Security Council members, the United States did not request a vote for military intervention in Iraq.

Full text of Colin Powell's U. N. speech is at http://www.state.gov/secretar\7rm/2003/17300pf.htm .

Possible Options for Invading Iraq

Desert Storm Type Tactics

This "scenario" would be a replay of the 1991 Persian Gulf War. There would be a more intense air campaign with faster deployment of ground troops than in 1991. The main invasion would come from Kuwait with air strikes from warplancs based in neighboring countries. It would require about 250,000 troops.

[]The disadvantage of this option was that it required basing a large presence of Western forces in the area.

[]There might have been repercussions from Arab countries reluctant to join the coalition.

[]Since it would require a large build-up of troops and supplies, Saddam Hussein might have been able to launch a pre-emptive strike.

Baghdad First

This strategy called for taking the capital of Iraq, Baghdad, where the power and communications center of Saddam Hussein's government was located. Surprise was very important here. The aim was a swift strike that would cause a collapse from within. About 100,000 troops were needed.

[]A disadvantage was that fighting would primarily occur in the Baghdad region, mostly in urban areas, with high casualties.

[]Saddam Hussein's republican guard had been trained in urban warfare and would probably do much of the fighting.

Rolling War

Tactics here included large forces seizing and establishing bridgeheads in northern, western, and southern Iraq.

[]Dissident groups, deserters, etc., would then rally, pressuring Saddam Hussein's regime internally. Saddam Hussein's government would eventually collapse.

[]Kurds in the north and Shia Arabs in the south would play a role in this campaign.

[]There probably would not be a direct assault on Baghdad.

[]About 300,000 troops would participate in this war.

Continued on next page

Possible Options for Invading Iraq, Continued

Rolling War (continued)

[] The main disadvantage to this plan was that the Iraqi opposition was splintered; the allies could not be certain that the Kurds and Shias would fully' support this plan.

[] The Iraqi Government had experience in controlling minorities in these areas.

Coup or Exile

Two other options were:

[] A surprise coup, sponsored by the CIA, would seize key installations in Baghdad while U.S.-led air forces attacked military targets around Baghdad.

[] Some Arab countries tried to persuade Saddam Hussein and his family to go into exile, with promised shelter and safety.

This early interview covering various political, administrative and international angles gives good perspective on the coming war by a PBS correspondent http://www.pbs.org/wgbh/pagcs/frontlinc/shows/gunning/intcrvicws/sciolino.html .

(This page intentionally left blank.)

Operation *Iraqi Freedom*

Preparation for War

Selected Topics

The topics covered in this section are shown in the table below.

Some Basic Assumptions & Conclusions

United States and coalition forces had two important questions in relation to winning the military campaign in Iraq:

[]Could the Iraqis put up a strong resistance?

[]Could the United States react as quickly to changes on the battlefield as it thought it could?

Conclusions were:

[]Even if the Iraqis could resist effectively and the United States had problems with managing the battlefield, the United States still could win the war.

[]Even if there were resistance at first, the U.S. military could isolate southern Iraqi forces.

The United States could then use airpower to take out the northern Iraqi military. It could also isolate and overcome Baghdad without heavy casualties. This combination of mobility and air power would make U.S. success very likely.

Continued on next page

PossibleWorst Case

The United States couldn't guarantee the

- []Timing of civilian casualties
- [] Range of civilian casualties

If intense aerial attacks were necessary, it might take up to two months to be effective. Then a full ground assault would follow in the hottest time, summer.

Civilian casualties would be heavy throughout the country, including Baghdad. Chemical weapons use by Iraq might be a reality in this case too. Even then, the U. S. would probably win the war. The chief issue would be: Is an extended war-fighting scenario politically tenable?

Keep these points in mind as you continue reading: weighing the military and the political is key to understanding the military campaign and later reconstruction of Iraq.

Military Build-up

Jan. 7 — The U.S. military built up forces in the Arabian Gulf, but said that it was not a sign that war with Iraq was inevitable (Defense Secretary Donald Rumsfeld).

Jan. 14 — During a White House meeting, President Bush said he was tired of the continuing deception on the part of Iraqi dictator Saddam Hussein and saw no evidence that Hussein was disarming as required by U.N. Security Council Resolution 1441.

Feb. 8 — The Defense Department enlisted commercial airlines to transport troops and equipment as part of the buildup for possible war with Iraq, with the activation of the Civil Reserve Air Fleet.

March 7 —

- [] U.N. chief weapons inspector Hans Blix reported back to the U.N. Security Council saying Iraqi disarmament would take months.
- [] The United States and Britain presented a revised draft resolution to the Security Council giving Hussein an ultimatum to disarm by March 17 or face the possibility of war.

Continued on. next page

Preparation for War, Continued

Last Minute Diplomacy

March 11 —

[]The White House rejected an}' plan to delay by one month the new resolution's March 17 deadline for Iraq to disarm or face military action.

[] France and Russia said they would veto any U.N. resolution that implied war.

March 16 —

[] The leaders of Spain, Portugal, United States and Britain met in the Azores for a final attempt at diplomacy. This resulted in an announcement that the allies would not pursue a second U.N. resolution.

[]U.N. observers evacuated the demilitarized zone along the border between Iraq and Kuwait ahead of the possible war.

The Ultimatum

March 17 —

[] Bush delivered an ultimatum that Saddam Hussein and his sons have 48 hours to leave Iraq.

[] Arms inspectors evacuated the country.

Operation *Iraqi Freedom:* The War Begins

"Decapitation" Attack

March 19 — The war begins. In a "decapitation" attack, a suspected gathering of Iraqi leaders, including Saddam Hussein, in Baghdad was struck by cruise missiles and bombs dropped by F-l 17 Nighthawk stealth aircraft. The aim was to kill Saddam Hussein in one of his palace bunkers. The attack did not succeed.

Airpower Focus

[] Coalition air forces also struck Iraqi long-range artillery emplacements, air defense sites and surfacc-to-surfacc missile sites.

[] Patriot PAC-3 missiles in Kuwait successfully intercepted two Iraqi missiles.

[] Coalition aircraft dropped almost 2 million leaflets over Iraq — the second million-plus leaflets dropped in two days. Coalition forces used the fliers and broadcasts into the country to get information to the Iraqi population.

Air Raids on Baghdad

Iraqi air defenses in Baghdad were at best sporadic during the coalition bombing of the capital

[]Iraqi anti-aircraft artillery (AAA) and surface-to-air missiles at times returning fire against coalition aircraft, and other times remaining silent.

[]Additionally, an important focus of the raids was Iraqi command and control centers and communications in the capital.

Reasons for Poor Iraqi Air Defenses

There were several possible reasons for the lack of AAA and surface-to-air missiles:

[]Coalition bombing was effective in degrading Baghdad's air defense system.

[] Coalition forces used B-2 and F-l 17 fighters, which eluded radar and could launch bombs from high altitudes, making most Iraqi air defenses totally ineffective.

[]Iraqi commanders were holding off on using AAA and surface-to-air missiles until they felt the final offensive against the capital was in progress.

Operation Iraqi Freedom: The War Begins, Continued

Degradation of Iraqi Air Defenses

Throughout the early air bombardment phase of the war, coalition forces steadily bombed Iraqi air defenses around the country, showing air supremacy over every aspect of Iraq except Baghdad.

As the air war progressed, witnesses inside Baghdad said that coalition planes began to fly low enough to be seen by those on the ground. This was the point at which coalition forces were certain that they had truly degraded Baghdad's air defense capability.

Ground War

On March 20, 2003, the ground war began.

[]The 3rd Infantry Division (ID) (Mechanized) rolled out of Kuwait into southern Iraq at 6 a.m. local time and met only slight resistance.

[]Special operation forces went into action throughout western and southern Iraq.

[]The 1st Marine Expeditionary Force (MEF) and the British crossed into Iraq to seize Iraq's oil fields near Basra.

Continued on next page

Operation Iraqi Freedom: The War Begins, Continued

Heading for Baghdad

[]By March 22, 2003, the 3rd ID had penetrated almost 100 miles into Iraq on its way towards Baghdad.

[] U.S. Marines had captured the port town of Umm Qasr. Umm Qasr, on the Kuwait border about 290 miles southeast of Baghdad, gave U.S. and British forces a port for military and humanitarian supplies and a base from which to clear out any Iraqi resistance in the south.

[]Coalition forces launched a massive missile assault on Baghdad. In the next 24 hours, coalition forces struck hundreds of Iraqi targets.

Map

The map below should orient you. It has the most important site names mentioned in this section on Operation *Iraqi Freedom.* Wc will repeat it throughout this "chronology" as you need to have a "geographic" orientation when discussing any military campaign. A detailed chronology of the Iraqi war crisis is at http://www.infoplcasc.com/spot/iraqtimelinc2.html.

The map below comes from:
httrr//www lib ntexas edn/mans/eifl03/iran smf)3 pif

Continued on next page

Operation Iraqi Freedom: The War Begins, Continued

Navy and Marine Efforts

[] Navy SEALs and coalition special forces seized Iraq's major gas and oil terminals in the northern Persian Gulf and airfields in western Iraq.

[]British Royal Marines seized the Fao peninsula.

[] The British also focused their efforts in the South at Basra.

[]U.S. and Royal Marines captured Umm Qasr.

[]The first elements of the 101st Airborne Division (Air Assault) crossed into Iraq.

Most of the chronology of Operation *Iraqi Freedom,* including the war in the North, South and the West is based on reports from the site: http://www.stratfor.com.

Top Priority: Baghdad

The Road to Baghdad

By March 22, 2003 The 2nd Brigade of the 3rd ID had clashed with Iraqi troops, killing 45 of the enemy with no U.S. casualties.

Bypassing urban areas and resistance, the division penetrated 150 miles into Iraq, approximately halfway to Baghdad.

On March 24, 2003, elements of the 3rd ID reached Karbala, within 50 miles of Baghdad. In a furious sandstorm, the 3rd ID battled the Iraqi Republican Guard's Medina Division outside of Karbala.

On March 26, 2003 -

[] Two separate columns of Republican Guard vehicles heading south out of Baghdad were struck by coalition air forces.

[]Elements of the 3rd ID encircled the town of Najaf, killing approximately 1,000 enemy troops in a three-day battle.

[] The 3rd ID and the 101st paused to refit on March 27, 2003. Elements of the 3rd ID fought Iraqi irregulars at the town of Samawah.

[]Units of the 3rd ID pushed north to Hillah while the rest of the division advanced to within several miles of Karbala on March 30.

On April 2, 2003 -

[]The 3rd ID crossed the Euphrates River at Musayyib and was within 30 miles of Baghdad.

[]U.S. Marines closed in on the city from the southeast after crossing a canal and the Tigris River at Numaniyah.

Into Baghdad

On April 3, 2003 -

[]Part of the 3rd ID fought its way into Baghdad's international airport as another part came to within 10 miles of the city.

[]U.S. Marines came to within 15 miles of the city.

Continued on next page

Top Priority: Baghdad, continued

Taking Baghdad

On April 4, 5 -

[]After a tank battle, the 3rd ID took over the entire Baghdad international airport and began encircling the city.

[]Marines fought Iraqi soldiers in a fierce fight before advancing to the southeastern edge of Baghdad.

[]An armored column of the 3rd ID drove through southwestern Baghdad, encountering pockets of resistance, without any casualties.

April 6 -

[]U.S. military planes land at the Baghdad airport.

[]The 3rd ID made another sweep through Baghdad, advancing from the west.

Saddam Hussein's Residences

On April 7, 2003 -

[]Elements of the 3rd ID seized Saddam Hussein's Republican and Sijood presidential palaces.

[]U.S. Marines advanced further into eastern Baghdad.

Collapse of Baghdad: April 9, 2003

[]Soldiers of the 3rd ID continued to search out and destroy remaining pockets of Iraqi resistance in Baghdad.

[] U.S. Marines captured the Rashid military air base in eastern Baghdad.

All structured resistance in Baghdad collapsed and the regime's control was broken.

[]Baghdad residents came out onto the streets and toppled statues and destroyed pictures of Saddam Hussein and his Baath Party.

[]The White House called this an "historic moment," but said the war had not yet ended.

[]Saddam Hussein's whereabouts were still unknown.

War in the South

On the Road to Baghdad

As mentioned earlier, the movement to Baghdad came from the south out of Kuwait. This meant that coalition forces had to eventually take the towns from Basra through Nasiriyah and Karbala. Much of this was covered in earlier sections Operation *IRAQI FREEDOM:* The War Begins and in Top Priority: Baghdad. In this section you will cover a little more about key regions of the south and their importance.

On March 21, 2003, U.S. Marines captured the strategic port in the southern Iraqi city of Umm Qasr, which is located along the Iraq-Kuwaiti border, about 300 miles southeast of Baghdad. Capturing this town gave U.S. and British forces a port for delivering military and humanitarian supplies, potentially free from Iraqi resistance.

Continued on next page

War in the South, Continued

Basra (Basrah)

The opening scenario of Operation *Iraqi Freedom* saw Marine expeditionary forces and Army special forces occupying the Basra region because Saddam Hussein's forces had set fire to a number of oil wells in this oil-laden region of Iraq.

[]Although Basra is a city smaller than Baghdad, it could have presented a stumbling block to coalition forces advancing on Baghdad. British forces would have rapidly been "depleted" by extensive house-to-house urban combat.

[]Any long-term siege of the city, with bombardment, would have meant starving the population, and would have been a true political liability for the coalition.

Basra was important, however, as part of the southern front. It lies on the north-south axis to Baghdad.

Operation *Iraqi Freedom:* War in the North

Kirkuk and Mosul: Keys to the North

Key to the northern front of Operation *Iraqi Freedom* were two important oil-producing centers: Kirkuk and Mosul,

[]If coalition forces could occupy these two population centers, the road from the north through Tikrit, Saddam Hussein's hometown region, to Baghdad would be secure. This would allow coalition forces a two-pronged approach, from the north and south, on Baghdad.

[]Besides coalition forces, which were much fewer than in the south, the chief player in this northern front were the Kurds and possibly Turkey, which bordered on Kurdish-dominated northern Iraq.

Continued on next page

Operation Iraqi Freedom: War in the North, Continued

Initial Phase

On March 26, 2003, the coalition opened the invasion of Iraq on the northern front. About 1,000 U.S. paratroopers landed in Kurdish-controlled Iraq.

Coalition warplanes bombed:

[] Iraqi positions near the town of Chamchamal in Kurdish-controlled northern Iraq. Chamchamal is about 20 miles east of the key oil city of Kirkuk,

[] Kirkuk in recent days. Aircraft also were heard flying over the nearby city of Sulaymaniyah.

Kurdish Involvement

Early on, Kurdish leaders hesitated in attacking major centers in northern Iraq; the Kurdish commander in the Dohuk area of northern Iraq said that:

[] There were too few U.S. troops in the North to open a new front against Iraqi military forces.

[] Air raids against Iraqi defensive lines during the early days of the offensive had not been very concentrated or intensive in the North.

Overrunning Iraqi Positions

Later it became clear that Iraqi forces in northern Iraq were not a strong match for Kurdish forces. With increased air support, Kurdish, together with U.S. ground forces, made steady progress against Iraqi forces.

A report on Kurdish militia fighting Iraqi forces at: http://www.washingtonpost.com/ac2/v\p-dyn?pagename=arlicle&node=&conlentId=A44821-2003Mar28&nolFound=true

Continued on next page

Mosul

On April 9, Mosul fell to essentially Kurdish forces supported by U.S. special operations forces and air power.

[]The Iraqis had withdrawn some of their divisions from Mosul, weakening its defenses substantially.

[]Mosul had limited strategic importance. Kirkuk, to the southwest, was more heavily defended. Kurdish troops would need increased coalition support to take it.

The fall of Mosul did clear the way to Baghdad. The road from the north to Baghdad passes through Tikrit, which is Saddam Hussein's home town and was his political base and supposedly a heavily defended stronghold

Kirkuk

On April 10, Kurdish and U.S. forces advanced to the western edge of Iraq's northern oilfields at Kirkuk and moved into the town of Dibis without encountering any Iraqi resistance. The oil facilities around the town appeared to be intact.

Later in the morning the Kurdish forces of Kurdistan Democratic Party and the Patriotic Union of Kurdistan occupied Kirkuk, which Baathist party militants and Iraqi forces had deserted earlier that day.

Report on Kurds taking Kirkuk at http://www.washingtonpost.com/ac2/wp-dvn?pagename=arlicle&node=&conlentId=A5395-2003Apii 0&notFound=true

Continued on next page

Operation Iraqi Freedom: War in the North, Continued

Tikrit

Because of continued pressure by coalition special operations forces and Kurdish forces in the north, Iraqi forces found it very difficult to withdraw to Tikrit and Baghdad. Tikrit was taken by coalition forces the same way Baghdad and Kirkuk had been captured earlier, with most of Saddam Hussein's forces disappearing before coalition forces had arrived.

- About 20 U.S. tanks entered the center of Tikrit along the main road from Kirkuk and occupied the city square.
- Earl}' on April 14, 2003, a task force of the 1st Marine Expeditionary Force, supported by Cobra helicopters and F-18s, attacked Republican Guard units on the southern edge of Tikrit.

War in the West

Purpose

The western front was a means to control airfields and to suppress any possible threatening missile systems there.

Special Operations

U.S. special operations forces gained control of western Iraq, from the Jordanian border, 170 miles to the Mudaysis airfield.

[]The area is mostly unpopulated desert but has a few airfields and areas that might have been used by the Iraqi military to hide SCUD missile launchers.

[]Coalition special operations forces, including British and Australian commandos, had already captured the airfields at the H2 and H3 pumping stations and were said to have been involved in skirmishes at Ar Rutbah and Akashat.

Continued on next page

War in the West, continued

Key Routes

The area reportedly under coalition control also included two important highways into Iraq:

- []The first, from the Jordanian border through Ar Ramadi to Baghdad.
- []The second, from Arar, in Saudi Arabia, through Karbala to Baghdad.

The Turkish "Card"

A Complex Challenge

In contrast to the direct attack on Baghdad, the war on the northern front had been more complex in terms of the

[]Local population

[]Effect on Iraq's neighbor, Turkey

Here coalition forces had to consider a number of variables:

[] Chiefly: the Kurds, indigenous to northern Iraq
[]Possible military action that Turkey might take in northern Iraq

Between these two "allies" as you saw in an earlier section of this handbook, is much tension, ethnic enmity, and nationalistic sentiment. These problems will play a part of the eventual solution to peace in northern Iraq.

Turkish Actions

Turkey took the following actions in relation to its interest in Operation *Iraqi Freedom:*

[]Turkey allowed U.S. aircraft to only use Turkish air space, but its parliament banned U.S. military planes deployed in the war against Iraq from taking off from or landing at Turkish bases, including the main air base in Incirlik.

[]Some Turkish trucks did assist coalition forces by carrying some supplies for them into northern Iraq.

[]Additionally, Turkish soldiers settled into Turkish villages and fields along the border with Iraq, positioning tanks, rocket launchers and artillery behind embankments.

Continued on next page

The U.S. Perspective on U.S - Turkish Relations

U.S.-Turkish relations probably outweigh many other international considerations in the Middle East. Regardless of Turkish "non-cooperation" during Operation *Iraqi Freedom* the United States needs Turkey.

[]Turkey is the key nation in the United States' geopolitical strategy in the region.

[]It influences events in the Balkans, the Middle East and in many of the Muslim countries of the former Soviet Union.

Importance of Turkey

Turkey is the single most important "piece of real estate" in the Euro-Southwestern Asian area. It has been a basic pivot of U.S. strategy since World War II.

A break with Turkey is inconceivable, regardless of how it treats the United States.

The Turkish Perspective: Importance of the U.S.

Turkey depends on the United States to

[] Guarantee its national security

[] Support Turkish armed forces in guaranteeing Turkish secularism

[] Prevent the disintegration of Turkey by controlling the Kurds, who are interested in a unified independent state, which would probably include parts of Iraq, Turkey and Iran

Continued on next page

The Kurds: Forcing a Symbiotic Relationship

The Kurds have cooperated with the U.S. forces in Operation *Iraqi Freedom* (the Turks did not). For this, the Kurds expect something in return.

[] The U.S. administration wants to reward the Kurds, but is limited by the Turks.

[] Any move toward a Kurdish sovereign state would destroy U.S.-Turkish relations, and the threat of an independent Kurdish state would be a nightmare to the Turks, who would see it spreading and influencing the Turkish Kurds of southeastern Turkey.

As you can see the Kurds are one of chief factors that keeps Turkey and the United States balanced in a dependent relationship.

The Turkoman Minority

To make the Turkish-Kurdish issue even more thornier, the Turkoman, a large Turkic minority living in Kurdish-controlled northern Iraq, are beginning to demand power in that region.

[] There have been clashes between Kurds and Turkomans in the Kirkuk area.

[] Naturally, Turks favor their "brothers" the Turkoman in any kind of "struggle" with the Kurds.

As you can see, ethnicity will continue to play a dominant role in any political stabilization of post-Saddam Hussein Iraq.

Statistics: Operation *Iraqi Freedom*

Length	Combat operations lasted officially March 20 - May 1, 2003
Official Names	[] U.S. - Operation *Iraqi Freedom* []United Kingdom - Operation *Telle.* [] Australia - Operation *Falconer*
Troops Deployed	More than 300,000 coalition troops were deployed to the Gulf region: [] 255,000 U.S. []45,000 British []2,000 Australian [] 200 Polish
U.S. Estimates	[] Casualties - 115 U.S.; 53 other coalition countries []Wounded - 2,110 []POWs - 8 (all rescued)
Iraqi Estimates	[] Iraqi troops - 350,000 []Combat deaths - 2,320 []Iraqi civilian deaths - 7,900-9,600 []Iraqi civilian wounded - 5,100 []Iraqi POWs - 7,300
Cost	$4 billion per month to keep troops stationed in Iraq Statistics from http://www.infoplease.com/ipa/A0908900.html

(This page intentionally left blank.)

After Operation *Iraqi Freedom*

Important Political Events in Post-Saddam Hussein Iraq

Selected Topics

The topics covered in this section are shown in the table below.

Administrative Structure

A civilian administrator, L. Paul Bremer, former ambassador and State Department counter-terrorism director, currently runs the U.S. administration of post-Saddam-Husscin Iraq. . (He replaced retired Lieutenant General Jay Garner, the first administrator.)

An excellent time line of post-Saddam Hussein administrative and managerial events is at http://www.cfr.org/reg index.php?id=613511 12.

The Most Wanted List

In April 2003, the United States developed a list of the 55 most-wanted members of Saddam Hussein's regime in the form of a deck of cards.

[] One of the earliest "members of this deck" taken into custody was Saddam Hussein's former deputy prime minister Tariq Aziz.

[] In May 2003, Saddam Hussein's sons Uday and Qusay Hussein were killed in a gun battle at a villa in Mosul.

[] In August 2003, Saddam Hussein's cousin, Ali Hassan al-Majid or "Chemical Ali," was captured.

[] In December 2003, in a night operation called Operation *Red Dawn,* U.S. forces captured Saddam Hussein in a village near Tikrit. He was hiding in an underground "spider hole" with a few weapons and $750k.

Rewards have been offered for information leading to the capture of the remaining "cards."

Continued on next page

Important Political Events in Post-Saddam Hussein Iraq, Continued

Early Administrative Measures	In May 2003, the UN Security Council approved a resolution [] Supporting the U.S-led administration in Iraq [] Lifting of economic sanctions The U.S. administrator in Iraq abolished [] The Baath Party [] The Iraqi army and police force as constituted under Saddam Hussein [] Institutions of Saddam Hussein's regime The interim governing council (IGC) of Iraq met for first time.
Contracting for Rebuilding Iraq	Two large U.S. corporations are participating in the reconstruction of Iraq. Halliburton Corporation was awarded a contract by the U.S. Agency for International Development for more than 1.7 billion dollars. A Halliburton subsidiary, Kellogg Brown and Root, was awarded a contract to [] Operate Iraqi oil fields [] Distribute oil from Iraqi fields Bechtel was also awarded a contract to rebuild bomb-damaged Iraq.

"New" Iraq, Old Neighbors

Syrian Challenge

A democratized Iraq could eventually present a potential threat to many Middle Eastern countries. An example is Syria, which could pose destabilizing problems for both coalition forces and the newly emerging post-Saddam Hussein Iraq.

[] Syria voted in favor of Security Council Resolution 1441 and maintained a behind-the-scenes dialogue with the United States despite condemning U.S. war plans.

[] Then it changed course, encouraging anti-war activity and sentiment. The president of Syria spoke out about what his country considers the U.S. potential for dominating the Middle East.

[] Initially, the country left its border with Iraq open so that anti-coalition voluntary forces could freely mass in Syria and launch "attacks" against coalition forces. Busloads of anti-coalition volunteers launched from Syria into Iraq.

[] Later, because of U.S. pressure and threats, the government closed the border though traffic in supplies to anti-coalition forces in Iraq continued.

Iran

As you learned earlier, Iraq and Iran have been at odds over the years in their efforts to dominate the Middle East. Additionally, the "favored" status of Iraq's Shias by the regime in Tehran has been a sore spot in the relationship between the two countries.

[] It is possible that the pro-Iranian Shias of Iraq could organize to carry out a guerilla style war against coalition forces. Iran has tremendous resources at its disposal to support this kind of a war, something that is uppermost in U.S. policymakers' minds.

[] Iran is at best unpredictable in its attitude and actions regarding the liberation and reconstruction of Iraq. While Syrian actions can be contained and are often very evident, Iran has more long-term goals, which in their complexity and subtlety, are difficult to assess.

New Iraqi Government

What Iraq needs: Peaceful Pluralism

Political and economic stability in post-Saddam Hussein depends on a government of all peoples of present-day Iraq.

[]The transition to political pluralism is not easy because Iraqis had lived more than three decades under Saddam Hussein's one-party dictatorial system.

[]A peaceful means for political parties to coexist, with respect to their differences, must be ensured. This political pluralism, which is one of the most elementary foundations of democracy, is a process that requires time to shape.

The Transfer Process

The aim of the transfer of power from the U.S. and coalition administration to the newly formed Iraqi government is a progressive shift of power, together with the consolidation of new Iraqi political institutions. Such a process requires time.

Establishing the ICC

The 25-member Interim Governing Council of Iraq was established in July 2003, based on negotiations between the

[]Coalition Provisional Authority
[]Former Iraqi anti-Saddam Hussein opposition parties

IGC Responsibilities

The council:

[]Appoints and dismisses ministers.
[]Controls the national budget.

[]Oversees the convening of a conference to draft a new Iraqi constitution.

General elections arc set to follow a referendum on this new constitution.

Continued on next page

New Iraqi Government, Continued

IGC Make-up

Most of those holding council scats arc

- []Members-in-exile from historically anti-Saddam Hussein groups.
- []Kurds from the liberated Kurdish areas of Iraq.
- [] Representatives of indigenous political parties some of which are at odds with the United States.

UN Approval

In August 2003, the U.N. passed a resolution

- [] Welcoming the IGC of Iraq to the U.N.
- [] Giving a 12-month mandate to the new U.N. assistance mission for Iraq

Council Action

- [] In early September 2003, the IGC appointed a cabinet of 25 ministers, reflecting Iraq's ethnic and religious composition.
- [] The cabinet was tasked to run the country's affairs until a new elected government was established.

Reaction to the IGC

Many Arab countries have not yet recognized the IGC as the government of post-Saddam Hussein Iraq because they consider the ICG a "puppet" of the United States.

France and other countries, which had been originally opposed to the invasion of Iraq, would like to see a more speedy transition of power to the Iraqi people.

Guerrilla-Style War

Low Intensity

CENTCOM has admitted that in its efforts to stabilize Iraq and to maintain the peace there, coalition forces in Iraq face a terrorist adversary using guerrilla style low intensity tactics.

Types of Opposition

The terrorist opposition is probably made up of the following types of individuals and groups or has attracted similar individuals

[] Remnants of military forces still loyal to Saddam Hussein
[]Remnants of the Baath Party
[]Sunni Muslims loyal to Saddam Hussein
[] Disgruntled Iraqi citizens who have lost patience with the U. S.-led coalition
[]Al Qaida (Al Qaeda) and other voluntary terrorist-type forces who have crossed over into Iraq to "liberate" the country from coalition forces

Terrorist Incidents: Examples

[] In, August 2003, the Jordanian Embassy in Baghdad was bombed, with 11 people killed. Twenty-two people were also killed when a terrorist bomb destroyed U.N. headquarters in Baghdad. A car bomb killed 125 people in Najaf, including the prominent Shia leader Ayatollah Mohammed Baqr-al-Hakim.

[] A female member of the Interim Governing Council was assassinated in September 2003. Dozens of people were killed when the Red Cross office in Baghdad was bombed.

Continued on next page

Guerrilla-Style War, Continued

Shift in Tactics?

[]There may be a change in the targets of terrorist attacks. Previously most of the targets had been U.S. personnel and officials associated directly with U.S. reconstruction efforts. Since late October 2003, more Iraqi citizens appear to be targeted.

[]The brutal suicide bombing of an Italian police base in Nasiriyah in November 2003 may also signal that terrorist forces have shifted their attacks to regions located outside the capital.

[]The murder of Japanese diplomats and Spanish civilians working in Iraq seemed to indicate that terrorists are targeting allied personnel in an effort to discourage coalition countries from sending support to U. S. forces in Iraq.

This discussion is based on http://ncws.bbc.co.uk/ 1/hi/world/middle cast/737483.stm.

Operation *Iron Hammer*

Operation *Iron Hammer:* A Response

To deal with the upsurge in terrorist activity (examples in the next section), coalition forces initiated Operation *Iron Hammer* in November 2003.

[]This operation was an aggressive attack on insurgents before they could strike coalition forces and coalition-friendly indigenous groups.

[]The coalition said it would use *"overwhelming force"* to suppress these attacks.

Terrorist Activities

[] Insurgents fired on a U.S. supply convoy north of Samara; U. S. forces returned fire.

[] Gunmen assassinated Hmud Kadhim, the Education Ministry's director general in Diwaniyah province in the southern town of Diwaniyah. This kind of assassination is part of the new terrorist policy to murder Iraqis who are working with coalition forces.

[]Two policemen were wounded when assailants tossed a grenade at a police station in Mosul.

[]A roadside bomb exploded in the southern city of Basra as a British civilian convoy was passing by, damaging a vehicle

Coalition Responses

[] A pair of 2,000-pound satellite-guided bombs was dropped near Baqouba, 30 „il,, northeast of Baghdad, on camps suspected of making bombs.

[] Near Kirkuk, fighter-bombers dropped 1,000-pound bombs on terrorist targets.

[] U.S. forces targeted an abandoned dye factory in southern Baghdad that was hit by artillery and air strikes.

[] U.S. forces also used heavy artillery, battle tanks, attack helicopters, F-16 fighter-bombers and AC-130 gun ships to take out targets in central Iraq, including Tikrit, Baqouba and Fallujah.

Details can be found at http://www.ctnow.com/news/custom/newsat3/sns-ap-iraq.0.7589745.storv?coll=hc-headlines-newsat3

Continued on next page

Operation *Iron Hammer,* Continued

Battle of Samara

This was the largest post-Saddam Hussein engagement for coalition forces. In l,te November 2003, coalition forces engaged insurgent forces, in a planned attack, in which scores of Fedayeen-like troops were routed or destroyed.

Capture of Saddam Hussein: Implications

With the capture of Saddam Hussein in December 2003, a number of key issues came to the forefront:

[] Would his capture decrease or increase the insurgent attacks on coalition forces and Iraqis working with coalition forces? The chief issue here was how involved in the insurgency "movement" was Saddam Hussein. From the way he was captured, hiding alone in an underground hole, coalition forces cannot be certain how much direction or command he had over such forces.

[]An important issue that has been a constant problem for the Bush Administration was Iraq's weapons of mass destruction. Hopefully, Saddam Hussein could reveal his plans and the locations of these weapons and missile systems.

[]If Saddam Hussein were, in fact, a key player in the insurgent terrorist attacks on coalition forces and Iraqis, he would be able to give critical information on the structure, location, strength, and leadership of these groups.

Reconstructing Iraq: Nation Building

Utilities and Sanitation

Although the water treatment system was antiquated and damage to the city power system also affected water treatment, about 60% of Baghdad now receives adequate water. The sewage system was not designed to support the 5.6 million inhabitants of metro-Baghdad. Because of looting and sabotage, untreated sewage is now deposited into the Tigris River.

[] Baghdad has electricity aboutl2 hours per day.

[] The water authority has restored the worst parts of the system.

[]U.S. Agency for International Development is in charge of the contract for equipment for water treatment.

[]The coalition has allocated more than $100 million for repairs to the sewage system.

Based on:
http://eraphics7.nvtimes.com/images/2003/11 /03/international/1 104HEAL. eif

Health

A health assessment of hospitals found that 65% of the equipment in the country was not working. In 2003, coalition allies are spending $422 million on health care. All 240 hospitals and 1,200 primary care clinics are now open.

Education

Since 1991, there had been little maintenance on school buildings. University and higher technical schools had been isolated from the international academic community. Improvements in education include the following:

[]School registration at the primary and secondary level has increased.

[]Teachers' salaries have increased.

[]Grants, in conjunction with U.S. universities, have been awarded to Iraqi centers of higher education to encourage study in the fields of archaeology, environmental research, agriculture, and legal education.

Continued on next page

Price of Reconstruction

[] About five U.S. soldiers have been killed per week since major combat ended.

[]The bounty paid to Iraqis who kill U.S. soldiers is $300.00 - 1,000.00.

[]Coalition allies have destroyed about 100 weapons and munitions per day.

[]The average price of a Kalashnikov rifle is now about $80.00. In May 2003, it was about $5.00.

Reconstruction: Eyewitness Report

This section is based on the article below from http://www.opendemocracy.net/debates/article-2-95-1573.jsp#. It is a first-hand account of some of the every-day "realities" of reconstruction in Baghdad.

Eight days in Iraq
by Yahia Said
6 - 11 - 2003

Key points made by the author, who spent eight days in present-day Iraq found "a people yearning for freedom, normality - an end to violence." He had been out of country for twenty-five years.

[]A "rediscovered humanity" - the courtesy and politeness of the local Iraqi police force would make inhabitants of Iraq's neighboring countries jealous: " T apologise for the inconvenience,' said the Iraqi policeman." The police force is the first government service organization in 35 years to work outside of Baathist control. Said feels this is a plus for the Iraqi people and to the detriment of Saddam Hussein.

[]"The wave of lawlessness which gripped the country upon the fall of the regime has largely subsided. Night curfews have been lifted and Baghdad residents are gradually venturing out of their homes. Many now eat out in outdoor cafes and restaurants, instead of hurriedly grabbing takeaways. Goldsmiths on Karrada Street, one of the few revived shopping destinations in the capital, displays its wares late into the night, a big change apparently from just weeks ago."

Continued on next page

Reconstructing Iraq: Nation Building, continued

Reconstruction: Eyewitness Report (continued)

[]Baghdad is alive with commercial activity. Government employees are now receiving salaries ten times higher than before the war. The streets arc clean. Children from poor families Said visited were eating bananas - something unheard of, under Saddam Hussein.

[]Many Iraqis forced into exile by Saddam Hussein's regime arc now returning to the city to set up political parties, start newspapers and to begin business enterprises.

[]Most dangerous is the ongoing political violence, according to Said. Most suicide bombers and terrorists target police stations, schools and international organizations. Many of these terrorists are considered foreigners.

[]"Some reports from the country seem to imply that Iraqis welcome at least some of the violence as legitimate resistance to occupation. During my eight days I did not meet a single person who shared this view. Indeed, the vast majority of people in Iraq - especially women, who represent 60% of the adult population - do not want the Americans to leave anytime soon.

[]"Iraqis want to regain control of their future and to become truly independent and free. They are sick of violence of all kinds, whatever the pretexts. They are tired of empty rhetoric and ideologies. Most Iraqis want to pick up the pieces of their shattered lives and move on. Whoever cares for them should help them do just that."

References

Sources

Besides the numerous web sites cited in this handbook, the list below of other sources may also help you:

Henderson, Simon. *Instant Empire: Saddam Hussein's Ambition for Iraq.* San Francisco: Mercury House, 1991. Henderson uses a balanced approach to discussing Saddam Hussein and his importance to the Arab and non-Arab world. He covers important subjects such as the family and political background of Saddam Hussein's rise; an analysis of opposition and corruption in Iraqi politics, and the geopolitical motives behind Saddam Hussein's invasion of Kuwait.

Karsh, Efraim and Inari Rautsi. *Saddam Hussein: A political Biography.* New York: Free Press, 1991. Karsh and Rautsi have written an analytical political biography of Saddam Hussein. The book "traces the meteoric transformation of an ardent nationalist and obscure Ba'th party member into an absolute dictator." The authors cover the personality of Saddam Hussein and how it affects his political actions.

Makariya, Kanan. *The Republic of Fear.* Berkeley: University of California Press, 1998. Makariya discusses Iraq as a one party state, the total control exercised first by the Baath Party and then by Saddam Hussein. The emphasis is on the security and secret forces of Iraq and the way they terrorize citizens.

Mylroie, Laurie. *Study of Revenge: The First World Trade Center Attack and Saddam Hussein's War against America.* Washington, DC: AEI Press, 2001. This is an extended analysis of the origins, background, planning, and eventual first bombing of the World Trade Center. The authors focus on the link between the bombers and Saddam Hussein.

Pollack, Kenneth. *The Threatening Storm: The Case for Invading Iraq.* New York: Random House, 2002. The author, one of the leading national security advisors during the Clinton administration, concludes that the post-Gulf War policy of containment against Iraq is no longer effective and believes that Saddam Hussein must be deposed by invasion.

Continued on next page

References, Continued

Sources, (continued)

Sciolino, Elaine. *The Outlaw State: Saddam Hussein's Quest for Power and the Gulf Crisis.* New York: Wiley, 1991.

Trcvan, Tim. *Saddam's Secrets: The Hunt for Iraq's Hidden Weapons.* London: Harper Collins, 1999. This book, by one of the leading United Nations weapons inspectors, focuses on how United Nations weapons inspectors uncovered Iraq's secret nuclear, chemical, and biological weapons programs and exposed the deceit that is at the root of Saddam Hussein's regime.

The site below has a chronology of the UN resolutions on disarmament inspections and Saddam Hussein's obstructionism. http://www.cia.gov7cia/publications/iraq wmd/Iraq Oct 20()2.htm#03.

The following is a list of online "bibliographical" sources that have extensive sites dealing with Iraq and Operation *Iraqi Freedom.*

http://www.au.af.mil/au/aul/bibs/iraq/iraq crisis.htm

http://www.au.af.mil/au/aul/bibs/tcrtod/tcrtod.htm

http://www.centcom.mil/opeiations/iraqi freedom/iraqifreedom.asp

http://www.lib.mich.edu/govdocs/usteiTOr.html

http://fas.oii: man dod-101 ops/index.html

http:/ newsbbc.co.uk/ 1/hi/world/americas/1584660.stm

http://www.ringnebula.com/Oil/Timeline.htm

http://www.cfr.org/reg index.php?id=6135lll

http: //www.army.mil/

Appendix A
Islam

(This page intentionally left blank.)

Islam: Basic Ideas

Introduction

You have probably heard the name *Islam* associated with Iraq. Most people in Iraq and in surrounding countries believe in Islam. In this section, you will learn about Islam, its main ideas, customs, and practices.

(http://thetruereligion.org/intro.htm - - a short introduction to Islam)

Definition of Islam

Islam means submission to the will of God in the Arabic language. The word *Islam* has the same root as the Arabic word *salaam,* which means peace. Muslims around the world often greet each other with the phrase "Salaam" in much the same manner as Jews say, "Shalom," as a greeting. Both salaam and shalom come from the same common language root, the three-letter root -- *s-l-m.*

(http://mcdia.isnet.org/off/Islam/basics/indcx.html) - discussion of various "doctrinal" aspects of Islam by noted theologians.

Allah

Islam is not named after a person like Christianity, which was named after Jesus Christ, Buddhism after Buddha, and Confucianism after Confucius. The central focus of Islam is always God.

The basic tenet of Islam is that you must submit to Allah (the Arabic word for God) and live according to His divinely inspired law.

The Precept

[] The key truth that Allah has revealed to mankind is that the only divine and worshipful "being" is Allah, the almighty God; thus all human beings should submit to and worship Allah.

[] Allah has 99 names, among them, The Gracious, The Merciful, The Beneficent, The Creator, The All-Knowing, The All-Wise, The Lard of the Universe, The First, The Last, and others.

No matter what "sect" of Islam a person belongs to, he believes in this important religious belief that is the common thread of Islam.

Arabs and Islam

Muslims (Moslems)

Now that you know that the religion is called Islam and the name of the God of Islam is Allah, what does the word *Muslim* mean? The word has the same s-l-m root as the words *Islam* and *salaam.* A Muslim is a person who submits to the will of Allah, regardless of race, nationality, or ethnic background. A Muslim has to submit completely and obediently to Allah and live according to Allah's message.

[]Muslims worship Allah alone and must not worship any person, place or thing other than Allah. They believe that Allah is the God for the Christians, the Jews, the Muslims, the Buddhists, the Hindus, and even atheists.

[]Orally repeating the basic belief of a Muslim found in the "motto" of Islam: *"Laa Elaaha illallaah"* which means, "There is no god but Allah," is the way a Muslim professes that he belongs to Islam.

Importance of Arabic in Islam

[] The "motto" above, *"Laa Elaaha illallaah,"* is in Arabic.

[]The prayers, sayings, and many blessings of Islam are all spoken in Arabic, just as the holy book of Islam, the Qur'an (Koran) is written in Arabic.

[]This common religious language of Islam is one of the ties that bind all Muslims together.

Continued on next page

Arabs and Islam, continued

Arabs and Non-Arab Muslims

Because of the importance of the Arabic language in Islam, many people assume that most Muslims are Arabs. This is not the case: About 80% of all Muslims are **not** Arabs.

[] In fact there are more Muslims in Indonesia than in the Arab Middle East.

[]Besides Indonesia, Muslims make up the majority in such non-Arabic countries as Turkey, Pakistan, Uzbekistan and other central Asian countries of the former Soviet Union.

[]There is a large minority in China too.

[]In Europe, Albania is Muslim. Bulgaria, Bosnia, and Macedonia have large Muslim populations. Elsewhere in Europe, there are immigrant communities of Muslims from Africa, Turkey, and Asia in France, Britain, and Germany.

[]In the Americas Muslims have increased in recent years, both from conversions and immigration; 20% of the population of Suriname in South America is Muslim. Today there arc about five million Muslims in America.

Muhammad (Mohammed)

Importance

Muhammad is ... the Messenger of God
and the last of the prophets ... (Qur'an, 33:40)

This one phrase from the Qur'an shows Muhammad's place in Islam and in world religion. Muslims believe that Muhammad is the prophet of Islam. His message is Islam. The revelation of Islam that Muhammad received is the *Qur'an* (Koran).

Muhammad and Other Religions

In terms of other world religions, Muhammad is the last prophet of God to mankind.

[] He is the last messenger of God.
[] His message applies to all mankind, regardless of their beliefs.
[] Muhammad is the successor to Moses, Jacob, Isaac, Abraham, and Jesus.

Muhammad is the prophet among all previous prophets and messengers.

[] He reinterpreted and corrected the message of Allah and gave mankind the true word from Allah. In a sense, Muslims consider that he "cleansed" the message.

[] In spite of his importance in Islam and for the world, Muslims do not worship him or ascribe any divine qualities to the prophet.

Muhammad, His Calling

Muhammad was born in Mecca (Makkah) in what is now Saudi Arabia, in the year 570 **CE.** His father died before he was born. Soon after his birth, his mother also died, so Muhammad was raised by his uncle.

Because of his trustworthiness and sincerity he was often asked to arbitrate disputes. He has been described as calm and meditative. Muhammad was very religious. It was his habit to meditate in a cave near Mecca.

At the age of 40, during one of his retreats, Muhammad received his first revelation from Allah through the Angel Gabriel. This revelation, which he continued to receive for twenty-three years, is known as the *Qur'an* (Koran).

Continued on next page

Muhammad (Mohammed), continued

Spreading Islam

As Muhammad began to preach the truth that Allah had revealed to him, he and his small group of followers suffered persecution.

[]This became so strong that in the year 622 CE Muhammad and his followers emigrated from Mecca.

[]This event, called the ***Hijra*** ('migration'), when they traveled about 260 miles to the north to the city of Medina, marks the beginning of the Muslim calendar.

[]After several years, Muhammad and his followers returned to Mecca, where they made peace with their enemies and established Islam.

[]Before Muhammad died at the age of 63, most of Arabia was Muslim, and within a century of his death, Islam had spread to Spain in the West and as far east as China.

Five Pillars of Islam

Introduction

The five tenets are the foundation for all Islamic belief. They are the cornerstones of Islam. (http://vvww.usc.edu/dcpt/MSA/fundamcntals/pillars/ - an introduction to the five pillars.)

Commitment

Called the Ash-Shahaadah, this is the "credo" of Islam:

"/ bear witness that there is none worthy of worship except Allah and I bear witness that Muhammad is His servant and messenger."

All Muslims subscribe to and repeat this belief.

Prayer

Called As-Salaat in Arabic, this is the requirement that adult Muslims pray at five specific times during a twenty-four hour day to Allah.

[]This is a ritual prayer requiring Muslims to say specific prayers and to coordinate hand and arm movements, eventually culminating in complete prostration or bowing down with the hands on the floor to show complete submission to Allah.

[]While performing As-Salaat, Muslims face in the direction of Mecca.

Alms

Known as the Az-Zakaat in Arabic, this means that Muslims must contribute to the support of those less fortunate Muslims—support of the poor. Muslims are encouraged to pay their Zakaat during the holy month of Ramadan.

Continued on next page

Five Pillars of Islam, Continued

Fasting

In Arabic this is called As-Sawam. It means that during the month of Ramadan, adult Muslims who are in good health

[]Cannot eat, drink, or smoke from sunrise until sunset every day of this holy month.

[]Must abstain from sexual relations.

The intention of fasting is to worship Allah. Ramadan is a sacred month because it commemorates the time when Allah first revealed the Qur'an to Muhammad. The time for Ramadan changes each year because it is based on a lunar calendar. Ramadan comes 11 days earlier each year.

Pilgrimage

Called Al- Hajj (Hadsh), this pilgrimage requires a Muslim in good health to travel to Mecca in Saudi Arabia at least once in a lifetime.

[]Specific preparation is necessary for the Hajj; for example there are rules about shaving and cutting hair and what to wear.

[]One of the high points of the Hajj is the *urnrah,* which means performing a ritual walk seven times around the sacred stone of Islam called the Kab'ah.

[]The Hajj commemorates the sacrifices and faith of Abraham, his second wife, Hagar, and their son, Ishmacl.

According to the Council on Islamic Education, it is the largest, regularly scheduled international gathering on Earth.

Al Qur'an (Koran) and Sunnah

Definition

Al Qur'an (the Qur'an) is the revealed word of Islam, the holy book of the Muslims.

- []Muhammad received the message in bits and pieces from the Angel Gabriel during a period of 23 years.
- []The word *Al Qur'an* means the recitation. During Muhammad's lifetime it was recited publicly.
- []The original language of the Qur'an is Arabic.

It is important to remember that Muslims believe the Qur'an is **the** word of Allah, untouched and uninterpreted by human beings. As Allah's final message to mankind, it is the holy book that supercedes all others: the *Old Testament,* the *Gospels,* etc.

Form and Content

The Qur'an is a collection of surahs or chapters, many of which you might call verses.

- []Muhammad received the surahs from Allah. Scribes selected by Muhammad usually wrote down these verses. Sometimes they wrote on wood, trees, parchment, and even stones. Many followers also memorized the Qur' an by heart.

- []In later years, the Qur'an was recopied and refined; accents and markings for reading were added.

Interpretation

- [] The Sunnah is the spoken word and acts of Muhammad. The Sunnah, Muhammad's actual words and deeds, uses the life of Muhammad to explain and expand on the verses and teachings of the Qur'an so that Islam can eventually become a world religion.

- [] A second form of interpretation is called hadith. This is narration about the life of the prophet or what he approved - as opposed to his life itself, which is the Sunnah.

The Qur'an, the Sunnah, and the ahadith (plural of hadith) are the basic body of Islamic religious doctrine for all Muslims, (http://www.afghan-network.net/Islam/) (http://www.usc.edu/dept/MSA/)

Jihad (Djihad)

Definition

The word *jihad* means struggle in Arabic. This means a struggle between the forces of Allah (good) and the forces of evil —an eternal struggle.

You have probably heard the word *jihad* in reference to Osama bin Laden and the Taliban. They consider the terrorism that they perpetuate a jihad. Actually, the word *jihad* has a couple of interpretations.

Self-defense

Islam allows fighting in self-defense and in defense of religion. There are rules of combat including prohibitions against harming civilians and against destroying crops, trees and livestock.

If good Muslims do not risk their lives in the cause of Islam, injustice would triumph. The Qur'an says:

> *" Fight in the cause of God against those who fight you, but do not transgress limits. God does not love transgressors."* (2.190)

War is the last resort and is subject to the precepts of Islamic sacred law.

Inner Struggle

The second and more personal meaning of jihad is an inner struggle. Each of us, according to Islam, is constantly fighting a "war" against self-centered desire, and egotism, with the final goal of attaining inner peace.

Islamic Schools (Sects)

Religious vice Secular Law: Western View

The United States and most other Western countries make a clear distinction between the rule of law and religion.

[]Religion is a separate personal issue that does **not** determine law.

[]A legal case may make use of religion, but religion is not the determining factor in most legal cases.

[]Simply put, a religious denomination, such as Methodism, Catholicism, or Judaism, doesn't have the power to rule on a case of adultery in a civil court or on a case of fraud in a criminal court.

The United States has a secular system of law; religion plays no direct part in such a system.

Islamic View

If you look at Islam, the perspective is different. Islamic law consists of guidelines and rules that determine all aspects of a Muslim's life from how to perform ritual prayer to conducting business transactions.

[]It also includes crimes and the appropriate punishments for each.

[]These Islamic laws arc based primarily on the Qur'an and are called the Sharia.

[]Muslims must apply these precepts handed down to Muhammad in the sixth century to present-day situations.

Interpretation and Re-interpretation

This kind of religious legal system, particularly how to interpret the Qur'an and the ahadith, has resulted in a certain degree of divisiveness in Islam, just ,s varied interpretations of Christian doctrine have caused Christianity to split into various denominations. The two main sects of Islam, which are the schools of Islam, are the

[] Sunni
[]Shi'ites (Shi'ia)

Continued on next page

Islamic Schools (Sects), Continued

Shi'ites (Shi'ia)

The basic divisiveness in Islam is based on an historical incident:

- After Muhammad died, there was confusion about who should succeed him as the leader of Islam. One group felt that the successor was Imam Ali who had been appointed by Muhammad according to Allah's decree. Ali was a relative of Muhammad and the first to accept Islam.
- This group, mostly from the household of the prophet, was a minority and became known as the Shi'ites. They also believe that the leader of Islam must be endowed with grace and benevolence and should be infallible.

Sunnis

The majority group, called Sunni Muslims, believe that:

- Muhammad did not choose a specific successor and probably assumed that after his death, Muslims would find their own leader.
- The prophet did not tell his followers how they ought to select their leaders or what qualifications their leaders should have.

The Sunnis chose a leader, later called the caliph, from outside Muhammad's household. The Shi'ites refused to accept the Sunnis' choice and split off from this main group.

Results of the Division

Because of this succession split, variations in Islamic doctrine, law, and practice have developed between these two main groups over the centuries.

- Sunni Muslims predominate in Saudi Arabia, Afghanistan, Pakistan, Turkey, and Indonesia.
- Shi'ites arc the majority in Iran and southern Iraq.

Sunnis make up about 83 percent of Muslims, according to the *Encyclopedia Britannica;* Shi'ites, about 16 percent; and a few other small groups, the remainder.

Osama (Usama) bin Laden and Islam

Bin Laden's Religious Background

Bin Laden's brand of Islam is based on extremely conservative Islam called Wahhabism, which was espoused by Muhammad bin Abd al-Wahhab (1703-87) H.... the founder of the sect and the co-founder of Saudi Arabia. The Wahhabis and the royal Saud family have long intermarried. The Saudis have adopted many of the conservative precepts of Wahhabism.

Wahhabism: The Way of a Muslim Life

Wahhabi Islam is a total religious system. It has an answer for every question about a Muslim's life including social, legal and spiritual aspects. Wahhabi Islam is ascetic.

[]Men should wear short robes and even avoid the black cords used on turbans and headgear.
[]Mosques should have no decoration.
[] Drinking alcohol is forbidden.

Punishment is based exclusively on the Qur'an, for example:

[]The right hand should be amputated for theft.
[]Adulterers should be stoned to death.
[] Murder and sexual deviation are punishable by beheading. To Saudi Arabia uses these punishments, especially beheading for capital crimes.

Bin Laden and the Taliban

The kind of radically, conservative state that the Taliban had developed in Afghanistan had roots in Wahhabism, which inspires and feeds on the Taliban's fundamentalist religiosity.

[]For Islamic fundamentalists, the Taliban have created what approaches a pure, ideal society based on Islamic religion and law.
[]This society fits bin Laden's mold too.

Continued on next page

Focus on Expansion and Terrorism

- While bin Laden follows the conservative Wahhabi tenets, what he most fervently supports is that his brand of Islamic faith must expand. (To support this kind of expansion, Saudi Arabia supported the Afghan *mujahiditi* when they were fighting the Communists. So did Osama bin Laden.) http://www.newvorker.com/FROM THE ARCHIVE/ARCHIVES/7010924 fr archivc03 - good article on bin Laden, his life and religion

- Now bin Laden regards the struggle as a war between two civilizations: His Islam and the non-Muslim civilization, specifically the United States. Toward winning this struggle, bin Laden has committed his forces to terrorism.

http://www.brookings.Org/fp/proiccts/tcrrorism/faqs.htm#qb4 - a set of informative questions about terrorism.

Bin Laden-sponsored Terrorism

- Besides the infamous 9/11 terrorist attacks against the Pentagon and the twin towers, bin Laden allegedly sponsored the attacks on American embassies in Africa.

- U.S. and Russian officials are also concerned that he is financing Chechen rebel operations out of Dagestan, a former Soviet republic in Asia.

A primary concern is that bin Laden's forces arc attempting to acquire or build chemical and/or biological weapons to use in their terrorist attacks. In the case of the Chechen rebels, he may be trying to help them acquire radiological dispersal devices.

Continued on next page

Osama (Usama) bin Laden and Islam, Continued

Al Qaida (Al Qaeda)

Al Qaida (Al Qaeda) is the name of the network of Islamic extremists that bin Laden has at his command to carry out his radical Islamic terrorism.

- Al Qaida consists of a group of about 3,000 commanders. Troops of Afghanis and Pakistanis number 200,000+.
- There are a number of terrorist centers, with numerous cells, in North America, Yemen, Saudi Arabia, Albania, Kosovo, Algeria, Chechenya, all the former Soviet republics of Central Asia, the Philippines, Egypt, Ethiopia and Somalia.

Bin Laden's "Army"

Al Qaida is bin Laden's personal "army." Bin Laden could probably rally an army of over 100,000+ men around the world, excluding those currently "in service" in Pakistan and Afghanistan.

- Many members of his overseas forces arc ready for both sustained and one-time operations.
- In the interim, the members of these terrorist forces return to normal "civilian" life in host countries. They are also available for terrorist operations on their own countries.

Funding for Al Qaida (Al Qaeda) Terrorism

The primary sources for funding Al Qaida-sponsored terrorism have been:

- The sale of drugs, primarily opium.
- Contributions from Saudi Arabia through various banks and from various Saudi businessmen.
- What might be called tribute money to Al Qaida from countries that do not want the organization to establish cells in their own countries.
- Funds from bin Laden's personal fortune.

Continued on. next page

Osama (Usama) bin Laden and Islam, Continued

Bin Laden and Saudi Arabia

Even though the country where he grew up, Saudi Arabia, is extremely conservative, bin Laden finds that it is not conservative enough. He thinks the following reasons, which are contrary to his purist concept of Islam, are grounds for overthrowing the Saudi royal family:

[]Alliance of the ruling Saudi family with the West

[]Saudi Arabia's dependence on American and other foreign troops who came to the country to defend it during the Gulf War

[] Corruption of the Saudi regime

(This page intentionally left blank.)

Appendix B
Quick Facts on Iraq

(This page intentionally left blank.)

Quick Facts on Iraq

Introduction

This short facts page covers a number of important subjects that will give you a clearer understanding of Iraq, Saddam Hussein, and the post-Saddam Hussein era. It highlights critical issues.

Deserts and Mountains

The Republic of Iraq located in southwest Asia, is bordered by Kuwait, the Persian Gulf, and Saudi Arabia on the south, Jordan and Syria on the west, Turkey on the north, and Iran on the east. The climate is dry, with cool winters and hot summer. Most Iraqis live in the area of the Tigris and Euphrates rivers. The north is mountainous; the south, desert-like.

Key Factors

Oil, which is the main export; Islam, particularly the division between Shias and Sunnis; and former Iraqi dictator, Saddam Hussein, are the key factors that have shaped Iraq today.

Size and Population

Area: 434,924 square kilometers or 167,924 square miles. Estimated population: @ 23,000,000.

Language and Ethnicity

About 80%+ are Arab. Chief language is Arabic. 23% of population is Kurdish, speaking a non-Arabic language; others are Turkish, Armenians, and Assyrians.

Religion 95% of people are Muslim. There are twice as many Shias as Sunnis. Small pockets of minority religions like Nestorian Christians.

Admin. Sectors

Country is divided into 18 provinces.

Continued on next

Quick Facts on Iraq, Continued

Population Centers	Principal towns include the largest, the capital Baghdad; the port town of Basra; important interior towns are Mosul and Kirkuk.
Government/ Legal System	The Interim Govern Council (IGS) of Iraq is tasked to draft a new constitution and to set up elections.
Politics	Various political parties drawn from indigenous groups and from returned exiles vying for power.
Economic Sectors	95% of all revenue was from oil exports. 40% of gross domestic product is from services; 6% from mining and manufacturing; 10% from agriculture, 30% of work force in agriculture. Dates and cotton are the biggest crops.
Additional Site	http://www.odci.gov/cia/publications/factbook/index.html — CIA site that has handbooks of countries of the world.

Appendix C
Chronology of Events, August 1990- August 2002

(This page intentionally left blank.)

Chronology of Events, August 1990- August 2002

1990

Aug 2	After months of tension, the Iraqi army invades Kuwait. The United Nations Security Council passes Resolution 660 condemning the Invasion and demanding Iraq's immediate and unconditional withdrawal.
Aug 3	Arab League calls for Iraq's withdrawal from Kuwait.
Aug 6	Council passes Resolution 661; imposing comprehensive sanctions on Iraq and establishes a committee (the 661 or Sanctions Committee) to monitor the sanctions.
Aug 12-15	Iraq offers two peace plans that are rejected by the United States.
Au 28	Jordan proposes a peace plan that is accepted by Iraq but rejected by the United States.
Sep 19	Morocco proposes a peace plan that is rejected by the United States.
Sep 24	France proposes a peace plan that is accepted by Iraq but rejected by the United States.
Nov 22	Most expert witnesses to U.S. Senate Armed Services Committee reject military option towards Iraq.
Nov 29	Security Council Resolution 678 authorizes use of force against Iraq if it has not withdrawn from Kuwait by 15 January 1991.
Nov 30	U.S. proposes talks; Iraq accepts.
Dec 5	CIA director, William Webster tells U.S. Congress that "economic sanctions and the embargo against Iraq ...have dealt a serious blow to the Iraq economy. In late November, Baghdad cut civilian rations for the second time since the rationing program began ...In addition, services ranging from medical care to sanitation have been curtailed." Although sanctions are hurting Iraq's civilian economy, they are affecting the Iraqi military only at the margins.

Continued on next page

Chronology of Events, August 1990- August 2002, Continued

1991

Jan 9	U.S.-Iraq talks fail.
Jan 13	U.N. Secretary-General's talks with Iraq fail.
Jan 16	Air war begins, destroying much of Iraq's civilian infrastructure.
Jan 21	Iran protests scale of bombing.
Jan 29	French defense minister Chcvcncmcnt resigns in protest against scale of bombing.
Feb 3	Pope John Paul II rejects the claim that the war against Iraq is a "just war."
Feb.28	War ends.
Mar 20	Ahtisaari Report to Security Council on humanitarian crisis in Iraq and Kuwait. "...Most means of modern life support have been destroyed or rendered tenuous." "Sanctions in respect of food supplies should immediately be removed." No remedy to humanitarian need, "without dealing with the underlying need for energy."
Apr 3	Resolution 687 begins cease-fire, establishes U.N. Special Commission on weapons, extends sanctions bv tying them to Iraq's weapons. U.K. ambassador Sir David Hannay states in the Council that "it will in fact prove impossible for Iraq to rejoin the community of civilized nations while Saddam Hussein remains in power."
Apr 5	Resolution 688 condemns "the repression of the Iraqi civilian population" in the ensuing civil war. Mid-Apr U..S, U.K. and France organize a "no-fly" zone in northern Iraq, while Operation Provide Comfort carves out an autonomous zone in a large part of the
Jul 17	U.N. mission to Iraq led by Sadruddin Aga Khan concludes that Iraq needs $22 billion that year to provide civilian services at pre-war levels. Aug 15 Resolution 706 acknowledges the Sadruddin Aga Khan Report and calls for oil sales not to exceed $1.6 billion over 6 months to be placed in escrow account, deducting 30% for a Compensation Commission, plus UNSCOM and other international obligations, leaving less than 1/3 of the Report's recommended amount for humanitarian aid.
Sep 19	Resolution 712 proposes that Iraq be allowed $1.6 billion oil sales over six months, of which $900 million would be available for civilian needs, disregarding the Secretary General's request that the cap be raised.

Continued on next page

Chronology of Events, August 1990- August 2002, Continued

1992

Feb 1	Iraq rejects 706 and 712.
Feb 5	Council declares that Iraq "therefore bears full responsibility for their humanitarian problems."
August	U.S., UK and France establish no-fly zone in southern Iraq

1993

Jan 13	U.S., U.K and France attack Iraq with aircraft and cruise missiles. U.S and U.K continue air strikes on January 17 and June 26.

1995

Jan	U.N. Secretary General Boutros Ghali issues a report calling sanctions a "blunt instrument"
Apr 14	Resolution 986 allows Iraqi government $2 billion in oil sales every six months. 13% of total available funds set aside for U.N. use in the northern governorales. Sanctions Committee must review and approve all supplies purchased through escrow account.

1996

May 12	U.S. Ambassador to the UN Madeleine Albright, in response to claims of half a million child deaths in sanctioned Iraq, replies: "I think this is a very hard choice, but the price - we think the price is worth it."
May 20	Iraq is no longer able to provide survival sustenance for its civilian population. Iraqi government and U.N. reach agreement on implementing Resolution 986.
Sep 3-4	In Operation Desert Strike, U.S. fires cruise missiles at Iraqi targets
Dec 10	First oil sales start, beginning the Oil-for-Food program. It has since been renewed mostly in six-month phases.

Continued on next page

Chronology of Events, August 1990- August 2002, Continued

1998	
Feb 20	Oil-for-Food oil sales cap increased to $5,256 billion per six month phase.
Dec	UNSCOM's credibility is undermined by evidence that staff members seconded to the agency by the United States have compromised the independence of the agency and engaged in espionage and covert action to overthrow the Iraq government.
Dec 15	U.N. weapons inspectors withdraw from Iraq due to impending aerial attacks by the United States and the UK.
1998	
Dec16-19	Operation Desert Fox air campaign by U.S. and U.K.
1999	
Mar 30	Security Council panel report finds that Iraq had "experienced a shift from relative affluence to massive poverty" and predicted that "the humanitarian situation in Iraq will continue to be a dire one in the absence of a sustained revival of the Iraqi economy, which in turn cannot be achieved solely through remedial humanitarian efforts."
Aug 12	UNICEF estimates that an additional half million children under five who would be alive under normal circumstances had died in Iraq between 1991 and 1998.
Dec 17	Resolution 1284 offers improvements in Oil-for-Food, although less than those recommended by the Security Council panel, and expresses its intention to suspend sanctions with the "fundamental objective of improving the humanitarian situation" in Iraq. The oil sales cap is removed and some items are allowed into Iraq with automatic Security Council approval.
2000	
Jun8	Resolution 1302 establishes a team of "independent experts to prepare by November 26, 2000 a comprehensive report and analysis of the humanitarian situation". Iraqi government does not allow the team to enter its territory. Security Council rejects the alternative of a report based on UN agency information and other reliable external sources.
Dec 5	Resolution 1330 further expands lists of humanitarian items. Compensation fund reduced to 25% from 30% of oil revenues with the additional resources targeted to vulnerable groups.

Continued on next page

Chronology of Events, August 1990- August 2002, Continued

2001

May-Jun	U.K., French, and Russian draft resolutions propose various new approaches. The U.K. proposes a Goods Review List of potential dual-use items and land-based border monitoring of Iraq trade. Objections by Russia and by Iraq, as well as differences among Permanent Members blocks Council action.
Jun 6	One-month extension of Oil-for-Food under previous conditions.
Jul 4	Lacking agreement with Iraq, five-month extension of existing Oil-for- Food.
Nov 29	Oil-for-Food program extended by six months in Resolution 1382. Resolution proposes a Goods Review List to be adopted in May.

2002

Jan 29	U.S. President George W. Bush declares Iraq to be part of an "axis of evil" in his State of the Union message to Congress. Thereafter, reports abound of plans for a large-scale U.S. military attack on Iraq.
Feb 26	OIP Director Benon Sevan warns the Security Council of a "financial crisis" in the humanitarian program due to the dispute over oil pricing.
May 14	Resolution 1409 adopts Goods Review List.
Aug 1	Iraq Foreign Minister Naji Sabri writes to U.N. Secretary General Kofi Annan suggesting that Iraq may be ready to allow arms inspectors back into Iraq, but scepticism remains that inspections will resume

The source of this chronology is at http;//www.globalpolicy.org/securily/sanction/iraql/2002/paper.htm#appl .

Made in United States
North Haven, CT
04 November 2026

21774386R10187